ADVANCED
WATCH AND CLOCK REPAIR

ADVANCED
WATCH AND CLOCK REPAIR

H. G. HARRIS

EMERSON BOOKS, INC.
BUCHANAN, NEW YORK
1973

Library of Congress Catalog Card Number 73–81498
STANDARD BOOK NUMBER 87523–181–0
MANUFACTURED IN THE UNITED STATES OF AMERICA

Contents

Preface

During my years in Hatton Garden, London, it was suggested that I write a book, not for the skilled watch or clock maker, or the professional man, but for the beginner. A book that was simply written and clearly illustrated that would capture the interest of men of all ages and give them a hobby that was new and different, or more ambitiously an introduction to an ancient craft that has developed into a science.

So it was that in 1961 HANDBOOK OF WATCH AND CLOCK REPAIRS made its first appearance in the bookshops, and has since enjoyed increasing popularity.

During the last ten years many of my readers have asked for advice and guidance on matters not covered by the first book and so it was decided that another book should be written that would include subjects and techniques that are a little more advanced.

In writing this book I have assumed that the reader has acquired a working knowledge of the principles of operation of a basic watch or clock, and that he is capable of dismantling and hand cleaning a simple movement.

For the help that was given to me in supplying information and drawings I wish to express my appreciation and thanks to:

Smiths Industries Ltd. (Clock and Watch Division) Great Britain
Timex Corporation, Arkansas, U.S.A.
L & R Manufacturing Company, New Jersey, U.S.A.
Bulova Watch Company Inc., Bienne, Switzerland
Bulova U.K. Ltd. London
Eterna S.A. Grenchen, Switzerland
Eterna Precision Watches (U.K.) Ltd. London
Louis Brandt et Frere S.A. (Omega), Bienne, Switzerland
Kienzle Uhrenfabriken GmbH, West Germany

Portescap Reno S.A. (Vibrograf), La Chaux-de-Fonds, Switzerland

and my special thanks to Mr. Richard Lonergan for his many hours of care in the preparation of illustrations.

Part One

GENERAL

I

The Workshop

THE MAJORITY OF MEN WHOSE OCCUPATION IS THE REPAIR OF timepieces usually have a little corner tucked away in their homes where they can give junior's watch some first-aid treatment or grandmother's old mantel clock a new lease on life.

With some, the practice of horology is no more than a hobby, but even so it can also be a profitable spare time occupation. There are two things to remember. Each job must be done well, and the charge must be reasonable. If these two rules are observed then it will soon become known that you do a good job and are fair with your prices, and people will come to you on recommendation.

The majority of homes can offer only limited workshop facilities but because we require so little space it is possible that with some thought and a little planning something can be done.

The ideal arrangement is to have a room set aside for this purpose only, as illustrated in Fig. 1. It is impossible to have too much daylight, and so plenty of window space is an advantage. When working in artificial light an adjustable bench lamp is essential and the room light should preferably be fluorescent.

Dust is to be avoided at all costs, so one should never have carpeting on the floor. Even a most practiced worker sometimes has an accident and drops a small part. The chances of ever finding it in a carpet are remote. Sheet PVC floor covering is best, particularly if it is light in color with little or no pattern. The floor is easily cleaned and lost parts are not difficult to find.

Some form of heating is required for cold weather and you may wish to consider the advisability of fitting a small electric extractor fan in one of the windows.

Fig. 1 The workshop

To begin with, you will need a bench, a stool, and some tools. Other things can be added later as, and when, they become necessary.

Some men work on a table. There is nothing wrong with this arrangement providing the table top is the correct height and the table is rigid. If the work surface is too low you will quickly suffer a back-ache. The average height is 3 ft. and you will need a stool to suit.

If you are a handy-man you may be able to design and make your own bench similar to the one shown in Fig. 2. It is a great advantage to cover the top with white laminated plastic sheet such as Formica. The sheet provides a good surface on which to work, fine parts will come to no harm, the white background shows up very small parts including the smallest screws, and the surface can be cleaned by wiping with a moist cloth. Usually a beginner finds he can accommodate all his tools and miscellaneous pieces in the drawers of the bench. Later on, when he begins to accumulate equipment, he will need a small tier of partitioned drawers or trays in which to keep his spare parts.

There are a number of ways in which this can be accomplished. One popular method is to make a narrow wood frame with shelves of hardboard or plywood and in each compartment place a rectangular tobacco tin, or a plastic box with lid that can be obtained from material dealers, or even a home-made cardboard box. The contents of each container can be shown on a stick-on label at the front.

Fig. 2 The workbench

Pieces of equipment not in regular use are best kept in a cupboard out of the way and safe from dust.

If you decide to subscribe to a horological journal, the issues will accumulate. From time to time you will probably write for catalogs from suppliers, and servicing notes and descriptive literature from manufacturers, and in no time at all you will find you have a reference library. Here again you will need cupboard space or possibly a shelf or two with book ends.

Do not hesitate to write to suppliers for a copy of their catalog.

Usually they are generously illustrated and for the beginner they are packed with a wealth of useful information. Most suppliers make a nominal charge for their catalogs and they are well worth the price.

The range of general purpose tools used in the watch and clock trades is very wide, and to this can be added the tools and pieces of equipment produced by watch and clock manufacturers for servicing and testing their own timepieces. It would be impossible to mention them all and many of them you may never need, and so in this chapter we will refer only to basic tools, and leave any specialist tools to the chapters concerned.

EYE LOUPES: (Fig. 3) These are available in single or multi lens models with focal lengths ranging from ¼ in. to 5 in. They are designed to fit into the eye socket but spring steel headbands are available to hold the loupe in position. For those who wear a loupe for any length of time a headband is an advantage.

Fig. 3 Eye loupes

There are also spectacle loupes for those who wear glasses. These loupes are clipped to the spectacle frame and are hinged so that they can be swung clear when not needed.

Most watch repairers have two loupes, and some have three or even more. A single lens with a focal length of 3 in. is a good size for general work, and a single 2 in. lens is better for close inspection work. For very fine detail such as the inspection of pivots and jewel holes a double lens loupe with a focal length of ¼ in. will be found to give best results.

SCREWDRIVERS: (Fig. 4) Watchmakers' screwdrivers are made in different sizes ranging from 0.50 mm to 2.50 mm and they are available in sets or singly with fixed blades or reversible blades. All have revolving heads. Those with reversible blades are fitted with threaded ferrules to enable the blade to be changed.

Many of the sets are supplied with a different colored top to each screwdriver to provide a quick and positive identification.

Stands are available into which the screwdrivers are placed vertically, ready for immediate use.

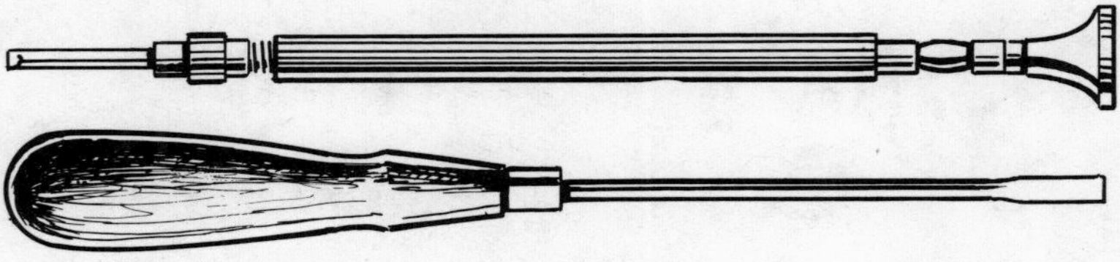

Fig. 4 Screwdrivers

Screwdriver blades need to be sharpened occasionally and sometimes they have to be reshaped. As explained in Chapter 2 the blades are hardened and tempered and they must not be allowed to get hot or the temper will be destroyed. This means that sharpening and reshaping cannot be carried out on a grindstone or emery wheel; all such work is done on an oilstone.

A correctly shaped blade will sit squarely in a screw slot, Fig. 5, and when turning pressure is applied, the screwdriver will move the screw without rising out of the slot and disfiguring the screw head.

The size of a screwdriver blade is in proportion to the size of the screw for which it is intended. The smaller the screw the more delicate is the touch needed to tighten it. With fine work the correct method of holding the screwdriver, once it is in position in the screw slot, is by the tip of the forefinger. Turning is done between the tips of the second finger and the thumb.

With the larger sizes of screwdriver, where a little more pressure is needed, the screwdriver head is positioned in the palm of the hand and thumb.

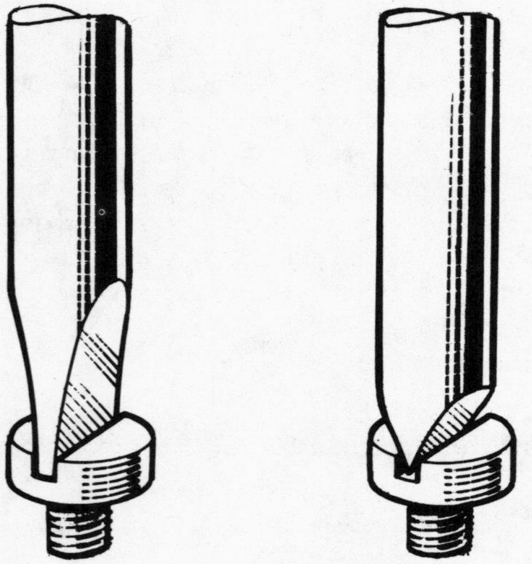

Fig. 5 Screwdriver blade

Keep screwdrivers away from anything magnetic. To work with a magnetized screwdriver is most frustrating.

TWEEZERS: (Fig. 6) The best tweezers are made from good quality carbon steel and finely ground to precision points. Great care is taken in the tempering because the points must be capable of gripping the smallest screw with safety. If the points are too hard they will snap when under tension. If the tweezers are soft or badly tempered the ends will open out at the points (Fig. 7), and the object being held can spring out and be lost or damaged. If tweezer points can pick up and hold under pressure a piece of hair then there is little wrong with the tweezers.

Take care of the points and above all do not allow the tweezers to fall to the floor. As with screwdrivers, keep tweezers away from anything magnetic.

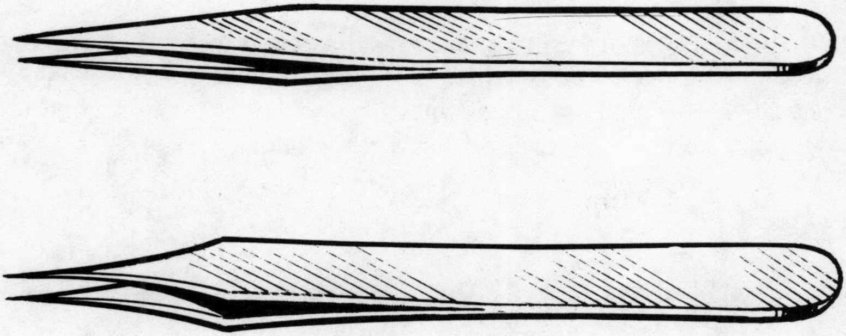

Fig. 6 Tweezers

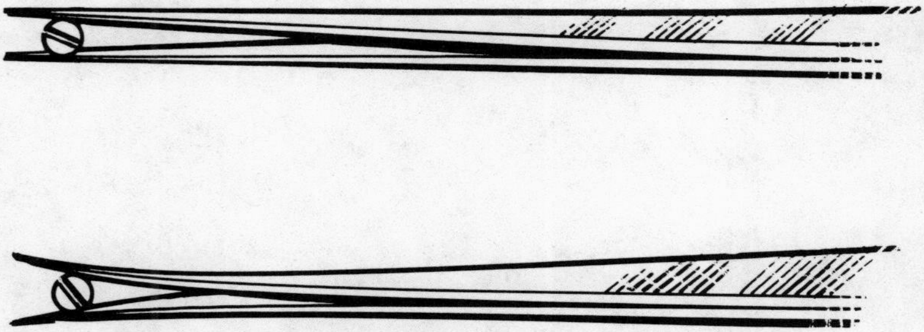

Fig. 7 Tweezers (TOP) correctly tempered
(BOTTOM) incorrectly tempered

OILERS: (Fig. 8) These are made from steel wire. One end is shaped as shown in the drawing, and the other end is fitted to a handle. To protect the end from damage and to keep it clean when not in use, a protective cap is provided.

Two sizes are needed for watches, one for general use and one for very fine work, and a larger size is required for general work on clock movements. Some manufacturers color the handles according to the blade size to provide easy identification.

OIL CUPS: (Fig. 9) These are usually polished boxwood containers carrying a colored glass cup and a close fitting cover. Two are needed, one for watch oil and the other for clock oil and they should be readily identifiable to prevent the possibility of using the wrong oil. One sure way of recognizing them is to have two different shapes or sizes.

An improvement is to make a solid wood base and cut out two shallow holes into which the boxwood containers can be placed, and a shallow trough in which to lay the oilers. A third hole can be cut to accommodate a pith holder.

Fig. 8 Oiler

PITH HOLDER: (Fig. 10) Oilers are cleaned by pushing the blade into a pith stick but this quickly breaks up the sticks and is wasteful. It is better to cut about 1 in. of metal tube approximately 1 in. diameter and pack it tight with lengths of pith stick and cut them off just above the level of the tube. The holder can then be inserted into a timber base for support.

A novel but very practical pith holder can be made from a brass cannon shell as shown in the drawing. Cut to length about 1 in. from the percussion cap. Remove the saw marks and burrs with a file and polish the edge with emery paper. Tightly pack with pith sticks, cut them to length, and stand the holder using the percussion cap as the base.

Fig. 9 Oil cups

Fig. 10 Pith holder

Before using the oil cups and oilers make sure that they are clean. Use the clock oiler to transfer oil from the bottle to the cup, but avoid as much as possible transferring more oil than will be needed for the job in hand.

Replace the cork in the oil bottle and store the bottle upright in case there is a slight leak in the cork. When the oil cup is not in use, the cover should always be replaced on the container.

Oil from the cup is transferred to the part being oiled, by using the appropriate oiler. It is necessary to only touch the tip of the oiler to the oil and a small quantity will be picked up. The oil will not pass back beyond the widest part of the blade.

The tip of the blade is now touched to the jewel hole or part to be lubricated, and capillary action will cause the oil to flow from the blade to the larger surface.

BLOWERS: (Fig. 11) Blowing into a movement to remove dirt or other pieces of foreign matter, or to dry out a part that has been washed, must be done with dry air. Blowing should never be

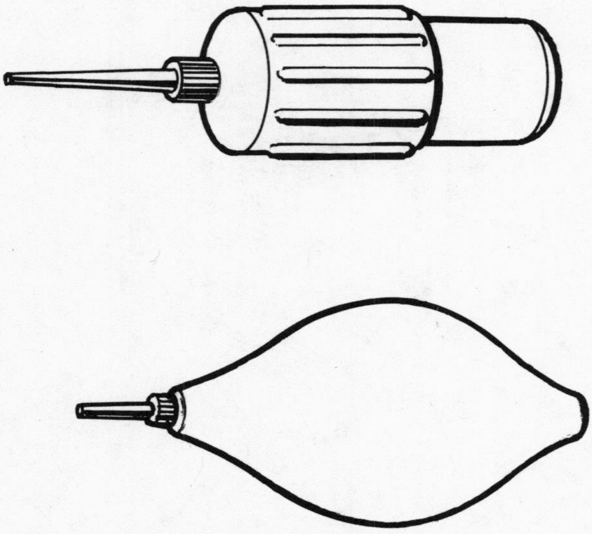

Fig. 11 **Blowers**

done by the mouth because breath includes minute droplets of moisture which, if deposited on the steel parts of a movement, will usually result in the formation of rust.

Whenever blowing has to be done always use a blower. There are two types in regular use. One is in the form of a rubber or plastic bulb which, when squeezed, forces air through a nozzle. The other type functions like a bicycle pump, in that a piston with a washer slides in a tube causing air to be pumped through a nozzle.

In both types the nozzle should be held between the tips of the forefinger and thumb and the squeezing or pumping action is done by the rest of the hand.

BRUSHES: (Fig. 12) After hand cleaning a movement the parts should be brushed to restore the original polish. A half hard brush is used for plates, bridges and the more robust pieces, and a soft brush is used on the more fragile parts.

Fig. 12 Cleaning brush

Have a stick of white chalk handy and stroke the brush across the chalk. This keeps the bristles clean and free from grease and acts as a polishing agent.

When cleaning French clocks and antique carriage clocks the lacquered brass work responds well to brush treatment. Make sure that the brush movements are made parallel to the graining on the surface of the metal, because otherwise the bristles may create the impression of scratching.

TRAYS: (Fig. 13) When disassembling a watch movement it is advisable to place the parts in a tray. These trays are partitioned so that the parts can be kept in groups; e.g. balance, escapement, gear train, winding, hand setting, etc., and close fitting transparent covers are provided to maintain cleanliness and to prevent damage and loss.

Fig. 13 Parts tray

MOVEMENT RESTS: (Fig. 14) A watch movement that is undergoing repair or overhaul needs to be supported. This is done by the use of rests that are made of boxwood or plastic as illustrated, or lengths of brass or aluminium tubing of different diameters.

The movement is placed on the correct size rest and held there

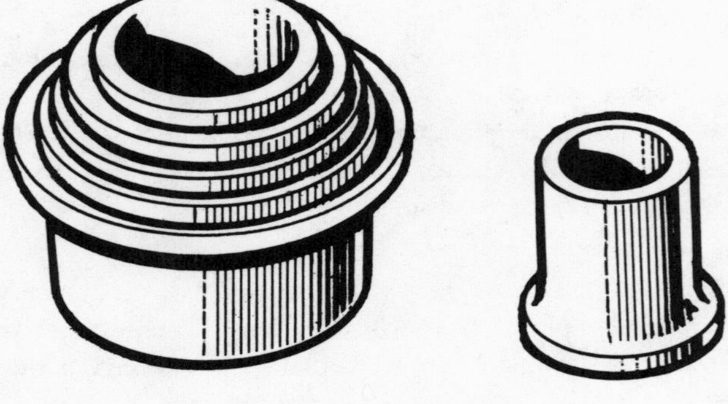

Fig. 14 Movement rests

by one hand, leaving the other hand free to operate. Work is done either with the rest on the bench, or with the movement and rest together held in one hand.

PLIERS (Fig. 15) Good quality watch and clock makers' pliers are accurately machined and carefully heat treated, and are usually completely polished. The flat nose and the chain nose pliers

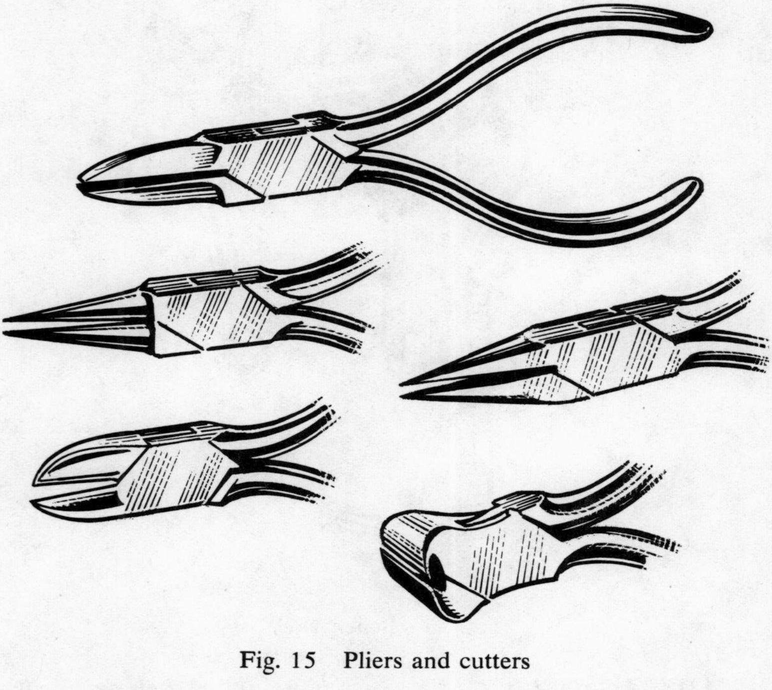

Fig. 15 Pliers and cutters

are available with corrugated jaws and smooth jaws. The corrugated are for general work, and the smooth are used to hold delicate and polished work.

BENCH VISE: (Fig. 16) There are two types of bench vises. One can be screwed to the bench as a permanent fixture, and the other is provided with a clamp enabling it to be fixed to the bench quickly, and just as quickly removed and put away. Of the two the most convenient is the portable model. Bench vises are made in different sizes measured by jaw width. The range is usually 1¼ in. thru 2½ in.

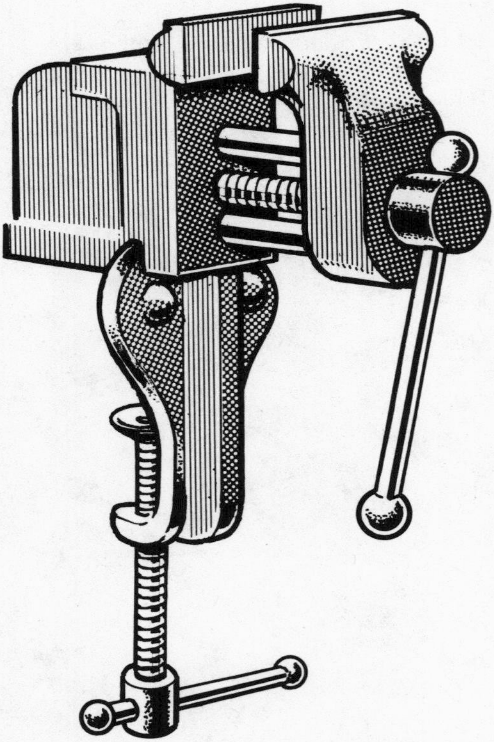

Fig. 16 Bench vise

HAMMERS: (Fig. 17) For general use a 2 oz. ball peen hammer is recommended, but for delicate work, and work involving the use of a staking tool, a staking hammer is needed.

FILES: (Fig. 18) There is a very wide range of files available to cover many aspects of work and only a tool catalog can hope to present to the reader the variety that is obtainable. There are many cross-sectional shapes, each with their own range of sizes and cutting grades.

The watch and clock repairer will concern himself more with the Swiss pattern than with the American pattern. The Swiss files are more accurately made because they are intended as finishing files for light precision engineering and instrumentation work, whereas the American pattern files are more suited to general engineering.

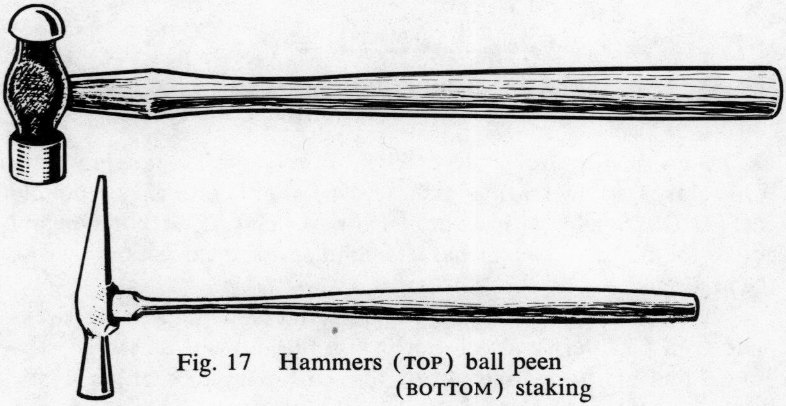

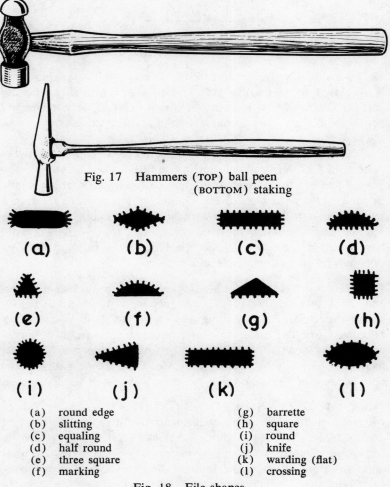

Fig. 17 Hammers (TOP) ball peen (BOTTOM) staking

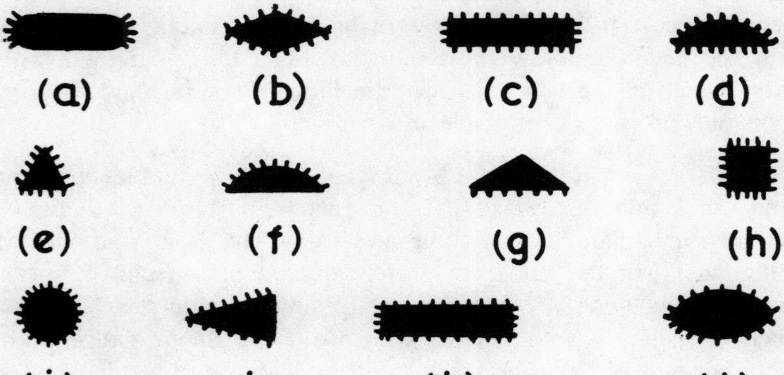

(a) round edge
(b) slitting
(c) equaling
(d) half round
(e) three square
(f) marking
(g) barrette
(h) square
(i) round
(j) knife
(k) warding (flat)
(l) crossing

Fig. 18 File shapes

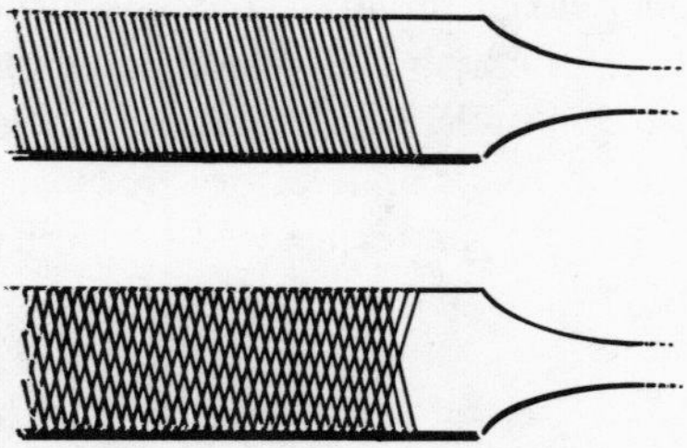

Fig. 19 Files (TOP) single cut
(BOTTOM) double cut

The Swiss files are made in eight cutting grades ranging from 00 (coarse) to 6 (fine) and these can be either single or double cut, Fig. 19. Single cut has parallel rows of cutting teeth diagonally across the file. Double cut has a second set of teeth passing across the first set.

The correct way to use a file and the method of cleaning is explained in Chapter 2. When not in use files should be stored separately and not tossed into a box one on top of another, as this is the quickest way to blunt them. One method is to keep them in a wooden rack drilled with a row of holes large enough for the file but too small to allow the handle to pass. Long narrow plastic bags which can be slipped over the files are useful, and the files can then be laid side by side in a tray.

DEGREE GAUGE: (Fig. 20) The gauge is simple in design and quick and reliable to use. Its purpose is to measure pinion heights, and diameters of staffs and stems etc. It is graduated in millimeters, or the European watch manufacturer's unit of measurement the ligne = 2.55883 mm, or approximately 3/32 of an inch. The gauge is sometimes known as a douziéme gauge probably because at one time the ligne was divided into 12 douziémes.

MICROMETERS AND SLIDING GAUGES: (Fig. 21) These two instruments are capable of measuring with great accuracy and are used extensively in the watch and clock industry.

Micrometers are made in two patterns, one for taking external dimensions and the other for internal dimensions, but we will consider only the external micrometer. The sliding gauge is capable of taking both dimensions.

While the instruments do not form part of a basic kit, their usefulness will become apparent when the repairer begins to make his own parts by hand or in the turns or on the lathe.

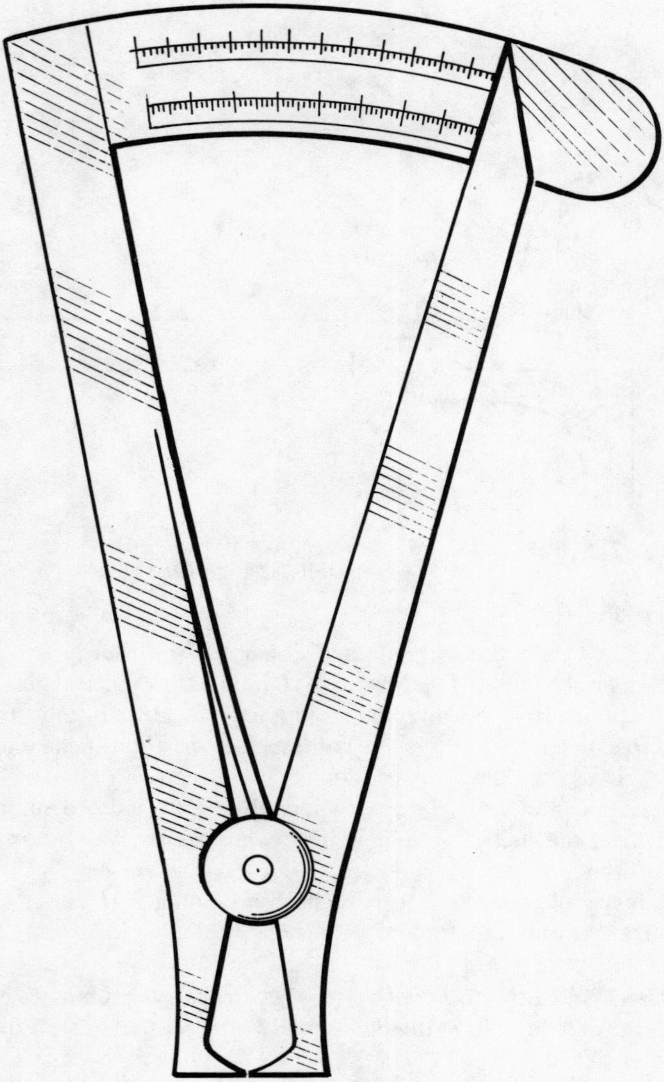

Fig. 20 Degree gauge

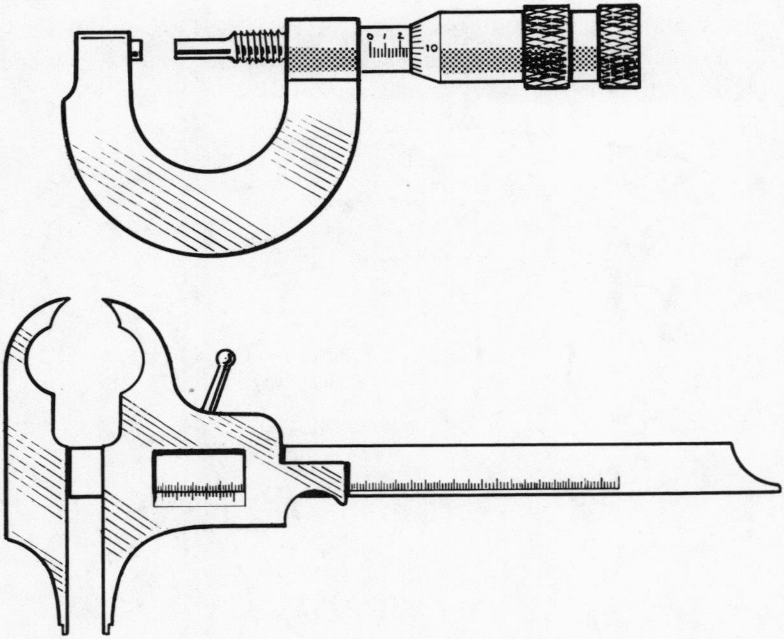

Fig. 21 Micrometer (TOP)
Vernier sliding gauge (BOTTOM)

The instruments are graduated either for the Metric System or the Imperial System. The Metric System is used by European countries and it has recentiy been adopted by British industry. In America the Imperial System continues to be the official measure but because a large proportion of imported watch and clock materials come from Europe, where they are made to metric dimensions, the metric system is in regular use by watch and clock repairers in America.

It is useful to be able to read both instruments in either system and so the four methods are given.

MICROMETER (Imperial): The sleeve is graduated over a distance of 1 in. and is divided equally into 40 parts, each division being 1/40 in.

The thimble rotates on a 40 t.p.i. (threads per inch) screw and each complete revolution will move the thimble along the sleeve a distance of 1/40 in. or 1 division on the sleeve.

The thimble is divided into 25 equal parts and therefore 1 division = 1/25 of a complete revolution.

If 1 revolution = 1/40 in. then 1/25 of a revolution = 1/40 × 1/25 = 1/1000 in. or 0.001 in.

Example 1: (*Fig. 22*)

sleeve reading	0.375 in.
thimble reading	0.023 in.
	0.398 in.

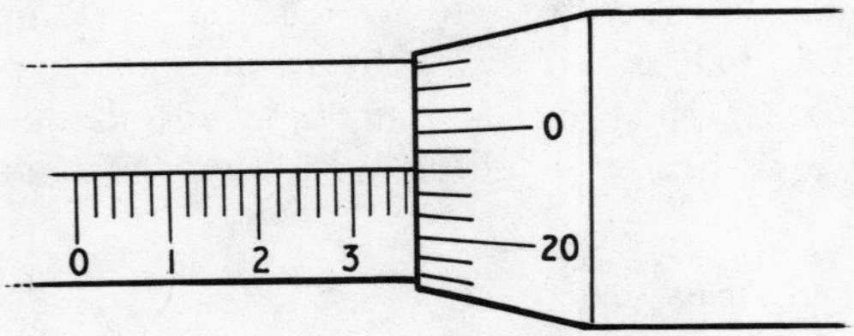

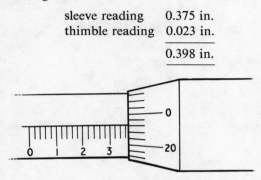

Fig. 22 Micrometer (imperial)

MICROMETER (Metric): The graduations on the sleeve extend for a distance of 25 mm. The datum line is divided into 5 equal divisions which are sub-divided into 5 equal small divisions each one being equal to 1 mm. These graduations are usually below the datum line. The graduations above the datum line mark the positions of ½ mms.

The thread on which the thimble rotates has a pitch of ½ mm (0.50 mm) which means that each time the thimble completes 1 turn it will have moved along the sleeve a distance of ½ mm.

The thimble is graduated in 50 equal divisions, and therefore if the thimble is turned 1 division the distance travelled by the thimble along the sleeve is $\frac{1}{50}$ of ½ mm = $\frac{1}{100}$ mm or 0.01 mm.

Example 2: (*Fig. 23*)

Below sleeve datum reading	= 18 × 1 mm	=	18.00 mm
Above sleeve datum reading	= 1 × 0.5 mm	=	0.50 mm
Thimble reading	= 21 × 0.01 mm	=	0.21 mm
			18.71 mm

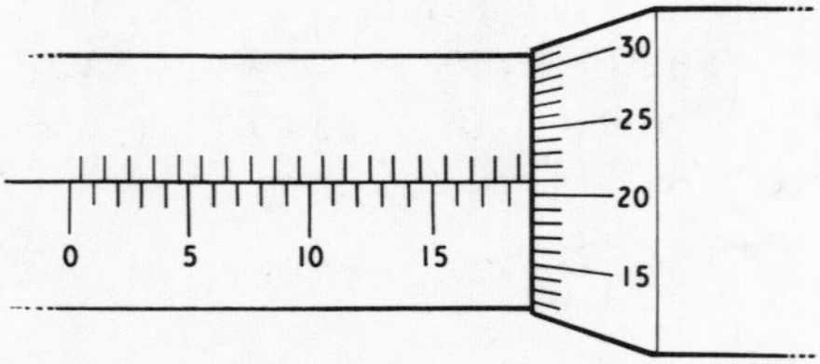

Fig. 23 Micrometer (metric)

SLIDING GAUGE (Imperial): Each inch on the main scale is divided into 10 equal divisions which are further divided into 4 equal divisions, where each small division is $\frac{1}{40}$ in. or $\frac{25}{1000}$ in.

The vernier scale is divided into 25 equal divisions and is equal in length to 24 small divisions on the main scale. Each division on the vernier scale must therefore be equal to 40 small divisions on the main scale divided by 25. This is expressed as:

$$\frac{24}{40} \times \frac{1}{25} = \frac{24}{1000}$$

The difference between 1 division on the main scale and 1 division on the vernier scale is $\frac{25}{1000} - \frac{24}{1000} = \frac{1}{1000}$ in.

Move the jaws up to the component being measured and take the main scale reading opposite the 0 on the vernier scale. First record the number of inches, then $\frac{1}{10}$ ths and then $\frac{1}{40}$ ths. To complete the reading, look along the vernier scale for a graduation that corresponds with a graduation on the main scale. This will indicate the number of $\frac{1}{1000}$ ths to be added to the first reading.

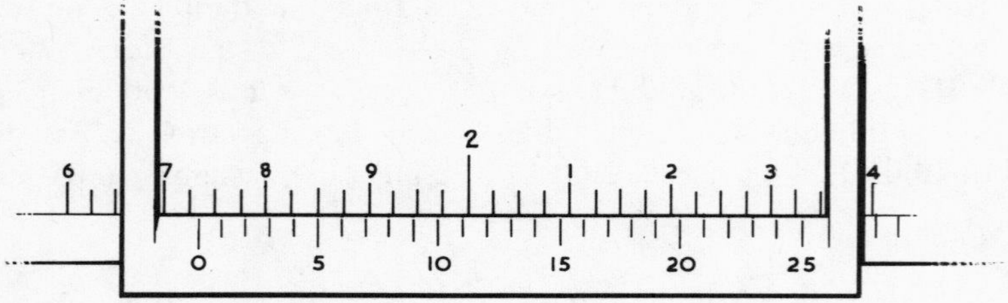

Fig. 24 Sliding gauge (imperial)

Example 3: (*Fig. 24*)

Reading on the main scale	= 1.725
Corresponding line on vernier scale	= .005
	1.730 in.

SLIDING GAUGE (Metric): The main scale is divided into millimeters. The vernier scale is divided into 10 equal divisions that are equal in total length to 9 mm on the main scale. Each vernier division is therefore equal to $9/10$ or 0.9 mm. This being so, the difference between one division on the main scale and one division on the vernier scale is $1/10$ or 0.10 mm.

When a measurement has been taken, read the number of millimeters on the main scale that are opposite the 0 on the vernier scale. Then find the graduation on the vernier scale that corresponds with a graduation on the main scale and add this figure to the first reading.

Example 4: (*Fig. 25*)

Reading on the main scale	= 4
Corresponding line on vernier scale	= 0.4
	4.4 mm

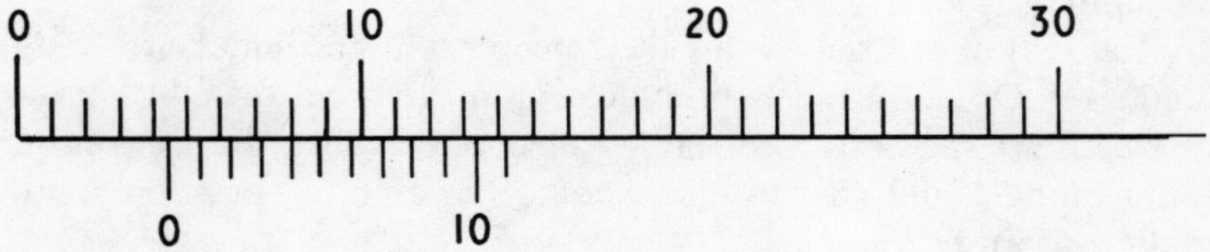

Fig. 25 Sliding gauge (metric)

II

Bench Craft

To obtain replacement parts for old watches and clocks is very difficult and usually impossible. The only way out of the trouble is to make the part oneself, or, if you do not possess the equipment needed for the job, the work will have to be done by an outworker.

There are many parts that can be shaped entirely with files and there is therefore no need for this type of work to be sent out. Sometimes there is the addition of a few drilled holes but this is simple enough. Filing plays an important part in the work of a watch and clock repairer and it is essential that he be able to file accurately.

The two most used metals the repairer will encounter are brass and steel. Do not use new files on steel in the belief that the sharpest teeth are the best. Steel can be filed better when the teeth have been blunted and so new files should be kept for brass and the older ones used on steel.

Normally, files are used dry but when working with aluminum it is an advantage to moisten the file with turpentine (white spirits) and it helps to prevent clogging of the teeth.

Similarly, when working with steel, particularly if a smooth file is being used, it is helpful to rub the file with a piece of chalk. This helps to prevent the steel filings from lodging between the teeth.

A good way to clean a file in order to rid it of metal particles, is to use the edge of a strip of brass as shown in Fig. 26. Never use a steel wire brush because it will ruin the file.

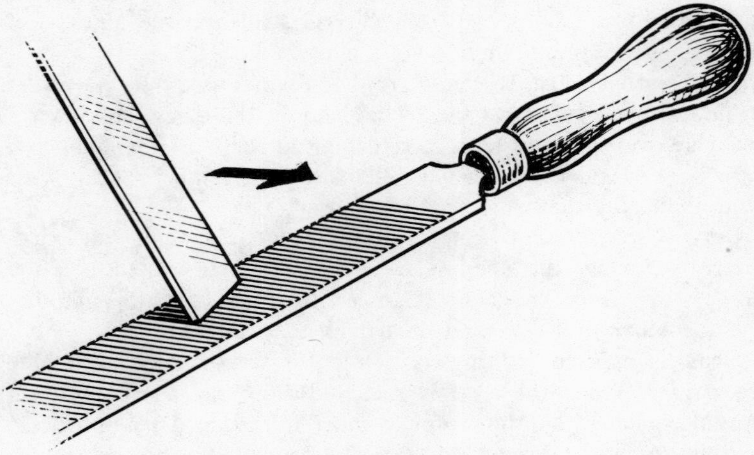

Fig. 26 Cleaning a file

The correct way to use a file on a large surface is by holding the forward end with the tip of the first finger and thumb of the left hand, and holding the handle in the right hand with the first finger outstretched along the edge of the file.

Place the file squarely on the work and make a definite but unhurried forward cut taking great care to keep the file horizontal. This is accomplished by beginning the cutting stroke with maximum downward pressure on the forward end of the file. As the file moves forward, this pressure is transferred so that when the file is halfway across the pressure is equal at each end. When the file is moved beyond this point the downward pressure continues to be transferred until, on completion of the cutting stroke, the maximum pressure is made at the handle end.

When filing a small surface the technique is the same but to a lesser degree.

The two basic steels one is most likely to find in the workshop are carbon steel and mild steel. Carbon steel is so called because it has a high carbon content which allows the metal to be hardened and tempered. The degree of temper (or toughness) that is introduced into the steel varies according to the temperature to which it is subjected.

All steel tools are made from carbon steel and the amount of temper depends on the function of the tool.

Steel springs must be made from tempered steel. If they were fashioned from soft steel they would adopt their new shape each time they were bent instead of returning to their original shape. If they were made of hardened steel with no temper at all they would snap the first time they were used.

Carbon steel cannot be tempered without first being hardened. After hardening, the steel is cleaned and then heated. As the temperature rises, the color changes to straw and through the reds to mauve and blue and then back to steel-color.

Straw is used for cutting tools, red for screwdrivers, and blue for springs. When the steel is removed from the source of heat the color change continues for a short duration. To compensate for this, the steel is removed from the heat a little before the selected color is reached.

Mild steel has a low carbon content and heating and cooling off has little or no effect; the metal remains soft.

Let us consider making a piece something like the setting lever spring that is illustrated in Fig. 91, item No. 56. We will need a strip of carbon steel slightly thicker than the broken part and with sufficient width for the job. Extra length is required for holding purposes during shaping and in this instance a piece about 4 in. long would be suitable. Make sure that the steel is soft because if it has been hardened and tempered it will be very difficult to cut and it will ruin the files. Test it by filing the edge. If the file bites and cuts into the metal then it can be assumed that the steel has little or no temper. If the file slides over the metal with no attempt at cutting, then it will be necessary to either soften the steel or select another piece. To soften carbon steel it must be heated to a cherry red and allowed to cool in air.

Having selected our piece of steel we now have to mark the shape of the piece we are going to make. Clean one end of the strip with smooth emery paper and then warm it slightly in a spirit flame. Rub a piece of beeswax on the metal so that the wax flows over the area to be marked, and then place the original pattern on the wax matching the two broken pieces together.

Hold the other end of the steel strip by means of an old pair of pliers and play the waxed end in the flame until the steel strip turns blue. Remove the steel from the flame and place it on a piece of cold metal to cool down. Then remove the pattern and it will be found that whilst the surrounding metal was turning blue, the area covered by the pattern remained unchanged and so a clear outline was produced in silhouette form, Fig. 27(a).

Clamp the steel strip in a bench vise and begin to file the unwanted metal from one corner, Fig. 27(b). The choice of files is governed by the shape of the finished piece. In this instance, flat, round and half round needle files would be a good choice.

When one corner has been filed to shape, repeat the process in the opposite corner, Fig. 27(c). Carry out frequent checks on the accuracy of the filing by placing the pattern on the work.

Continue with the shaping, Fig. 27(d), until the workpiece is almost complete, Fig. 27(e). A little more filing will now sever the new part from the steel strip.

Having completed the shaping, the holes are now drilled. Place the workpiece on a flat metal block and with a sharp centering punch, and a very light tap from a hammer, mark the centers of the holes to be drilled. Select the sizes of drills required by using the pattern as a drill gauge. Now place the work on a piece of flat brass plate, thick enough not to bend under drill pressure, and clamp the workpiece and brass in a bench vise fitted with brass chops or jaws, Fig. 28. The brass jaws will save the workpiece from being marked, and the brass backplate will give support to the work, and reduce the chances of the drill breaking when it has cut through the steel.

Twist drills are quite suitable for this operation and can be used in a spiral fluted hand drill that rotates forward and backward with an up and down movement of the handle. Remember that the twist drill cuts only in a clockwise direction. A hand wheel brace can also be used to hold the drill but for small fine work this tends to be rather heavy.

Having drilled the holes we can now file the work to the required

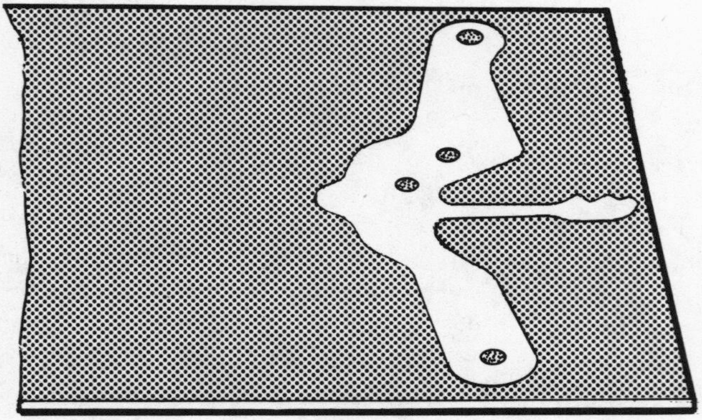

(a)

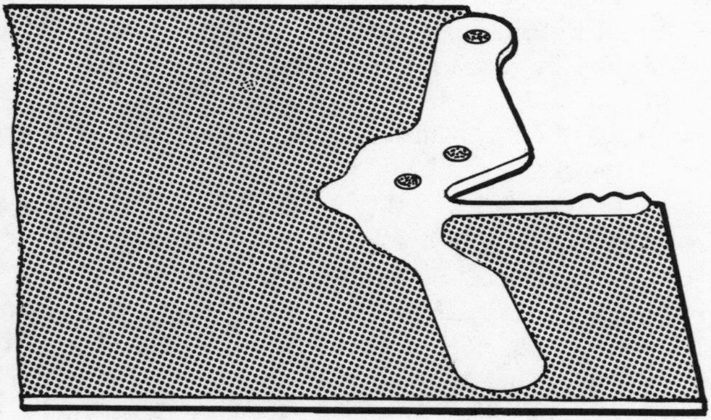

(b)

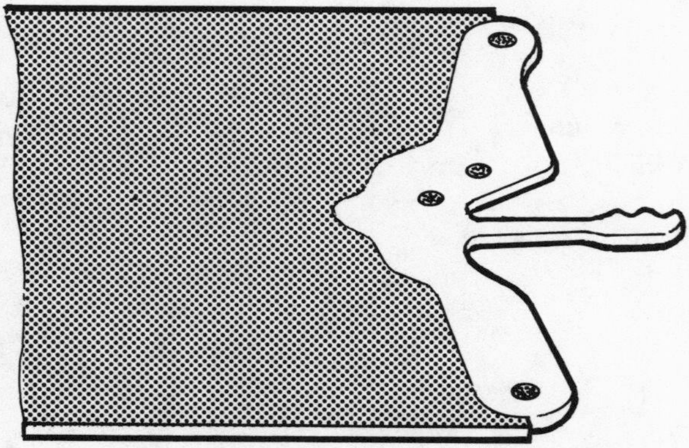

(c)

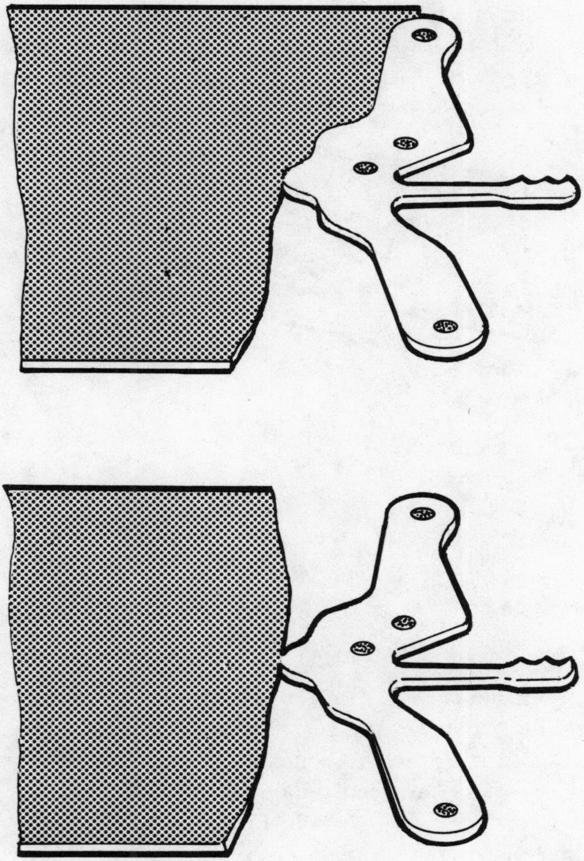

(d)

(e)

Fig. 27 Filing to shape

thickness. A small block of wood is needed with one face flat and smooth. Hold the wood in a bench vise with the smooth face uppermost and drive in a few small brass nails to prevent the workpiece from moving, Fig. 29. Cut off the nail heads and make sure that the nails have been driven below the level of the workpiece.

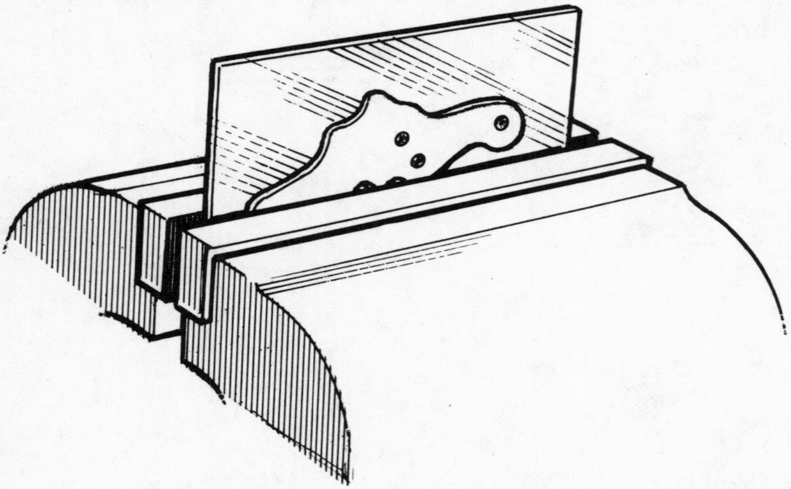

Fig. 28 Ready for drilling

Place a flat smooth file squarely on the workpiece and make a few slow and definite forward cuts. Lift the work from the pins and measure the thickness with a micrometer. Repeat the process until the required thickness has been obtained.

If countersunk screws are used to hold the piece in the movement the holes must be countersunk to receive them. Place the work in the bench vise with the brass plate behind, as for drilling, and with a bevelling tool countersink the holes. Sufficient metal must be removed to enable the screws to fit flush.

To give the new piece a finished appearance the file marks must be removed and a fine straight grain given to both faces. Place a piece of fine emery paper face up on a flat hard surface, such as a piece of glass, and carefully rub the work backwards and forwards on the emery paper by holding the work down with a stick of pegwood.

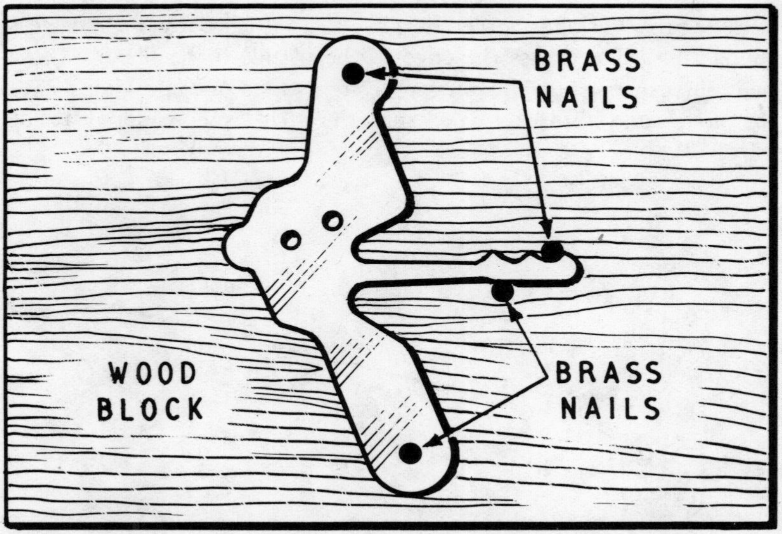

Fig. 29 Ready for filing thickness

Each time the hand set mechanism is operated, the locating pin in the pull out piece will deflect the center arm of the setting lever spring. If the new part is not hardened and tempered the center arm of the spring will suffer from metal fatigue and break off at its inner end.

The first process is to harden the metal. This is done by heating it to a cherry red color and plunging it into oil.

Obtain some fine soft iron binding wire from your material dealer and bind it round the workpiece in much the same way as an Egyptian mummy, Fig. 30. Pour some heavy grade car engine oil into a small tin. Wrap one end of a piece of stout iron wire around the workpiece as a handle and place the work in a bunsen burner or a methylated spirit flame. When the wire has turned to cherry red color quickly remove the work from the flame and immerse it in the oil where it must remain until cool.

If the workpiece was inserted into the flame without being wrapped its rate of heating would be so rapid that distortion would most probably take place.

When the work has cooled, remove the wire and test for hardness by passing a file across the edge of the metal. If the file bites, the heating process must be repeated. It is when the file slides over the metal and refuses to cut that the steel is considered to be hard.

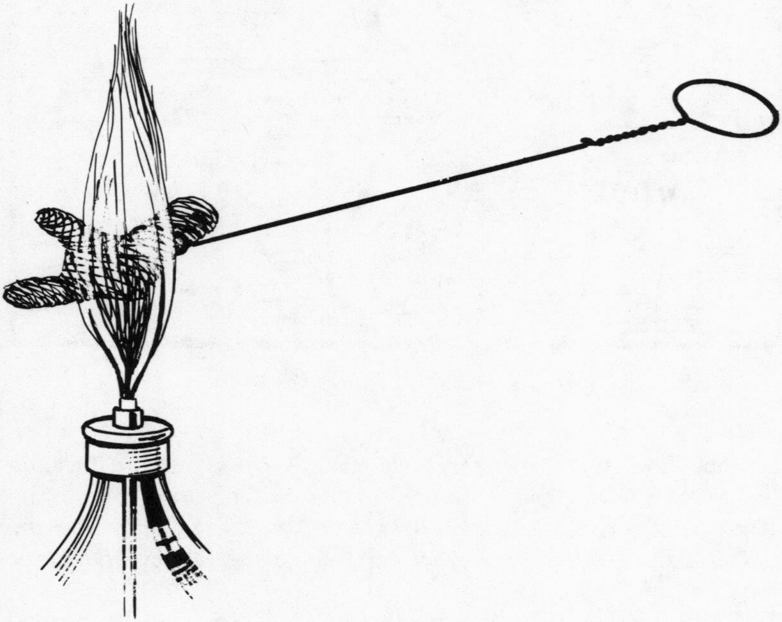

Fig. 30 Hardening process

The next process is to temper the steel and this can be done in one of two ways. One method is to take a small metal cup that will comfortably hold the workpiece, and around the cup wrap a length of iron wire for a handle.

Pour in some engine oil until the workpiece is covered and then hold the container in the flame until black smoke is given off.

Remove the container from the flame and allow to cool off. The steel will be tempered.

The other method of tempering is to clean from one face of the workpiece the oxidization caused by the hardening process. Rub the work on a piece of fine emery paper as before until the fine grained surface has been restored.

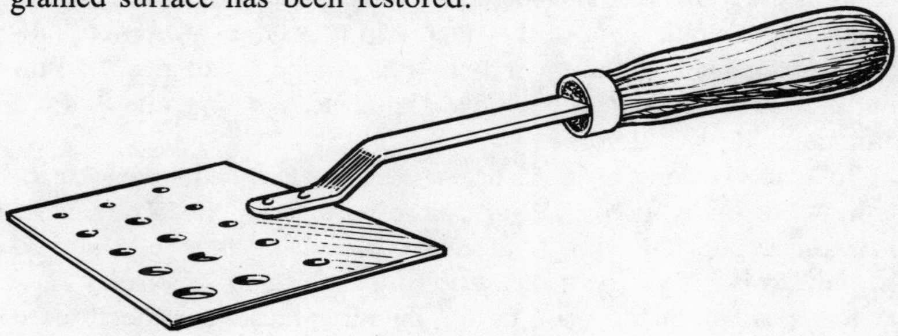

Fig. 31 Bluing pan

Place the work on a bluing pan, Fig. 31, and place the pan in the flame. Watch the work closely and as soon as the changing colors reach dark red into purple remove the pan. The tempering will continue for a few seconds more, during which time the color will change to blue.

If the bluing pan is left in the flame too long and the final color goes beyond blue into silver/grey, the temper process must be repeated.

If the work is removed too soon and the color stops at one of the reds, replace the work in the flame and withdraw it when the color starts to change.

When the heat treatment is complete the work must be cleaned of oxidization and bluing. Rub the faces on a sheet of fine emery paper as before, and draw file the edges. Draw filing is the action of using a file sideways and it produces a fine straight grain not obtainable when the file is used in a forward direction.

Bluing is frequently used as a means of decoration, particularly to hands and screw heads. The parts concerned must be well finished and then polished with diamantine. All traces of polishing paste and any grease must be brushed away and the work washed in cleaning fluid. The surface to be blued must not be touched by hand. Where there are finger marks there is grease and the steel will not blue properly.

To make a bluing pan we need a piece of metal sheet, brass, copper or steel, about 2½ in. x 2 in. x $\frac{1}{16}$ in. Drill a few rows of holes of different sizes and then rivet or screw a strip of steel to the plate. File the other end of the steel strip to a taper rather like that of a file tang. Heat the taper to a dull red heat and push it into the end of a wood handle. Give the handle a few taps with a hammer and drive it home.

Polishing is done by a paste on a hard surface. In the case of lathe work, or work in the turns, the paste is applied by a hand held metal polishing slip. If the work is flat and is to be polished by hand it is rubbed on a metal polishing block to which has been added a small quantity of paste. In either case the metal used depends on the material from which the workpiece is made. The rule is that the polishing slip or polishing block must be softer than the metal to be polished. The following chart is a guide to recommended combinations.

Metal to be polished

Hardened or Tempered Steel		Mild Steel		Brass, Gold, Silver	
Hand Work	Lathe Work	Hand Work	Lathe Work	Hand Work	Lathe Work
Plate glass (O) Zinc (D) Bell-metal (D)	Iron (O) (D) Mild Steel (O) (D)	Tin (D) Zinc (D)		Tin (D)	

O = suitable with oilstone powder
D = suitable with diamantine

The two most used polishing materials are oilstone powder and diamantine. As a substitute for oilstone powder a fine carborundum powder can be used if preferred. It is mixed and applied exactly as for oilstone powder.

The oilstone powder is mixed with clock oil into the consistency of cream. Its function is to grind rather than polish.

Diamantine is a fine white powder which is mixed with watch

oil. Do not mix too much at a time. A useful quantity is enough powder to cover a half cent piece, to which is added one drop of oil from an average size watchmaker's screwdriver.

Diamantine is used in the form of a very stiff paste; too much oil and the mix is useless. The mix must be turned over a few times and well pressed down with a knife blade before it is ready for use.

Two covered containers will be needed in which to keep the pastes. These containers are obtainable from any watch tool supplier and are provided for this purpose. They usually take the form of small round wooden trays into which are fitted flat metal discs that stand above the edge of the tray. It is on these metal discs that the powders and oils are mixed. The trays have recessed lids that provide dust proof coverage.

It is essential that the diamantine, both in powder form and when mixed, be kept scrupulously clean. Otherwise it will collect minute particles of foreign matter which are often of an abrasive nature. Their presence is indicated by fine scratches when polishing a flat surface.

Polishing slips for use on work held in the turns vary in shape and size. A good general purpose size for watch pinions, arbors, and straight pinions, is about 6 in. x ⅛ in. x ⅛ in. which is filed to shape at one end. Two useful shapes are shown in Fig. 32, and Fig. 33 illustrates typical polishers for balance staff pivots.

Many craftsmen prefer to use boxwood polishing slips instead of metal. These slips are available from tool suppliers in sizes ranging from ⅛ in. square to ¾ in. square and are suitable for use with oilstone powder and diamantine.

Arbors, pinions, pivots, etc., are best polished between centers in the turns. Both directions of rotation are used and the operation is quick and simple.

Start with a polishing slip charged with a small quantity of oilstone paste. It is the underneath surface at the front end of the slip that is used and this is touched to the mixed paste to pick up the small quantity that is needed.

Apply the slip to the work. Pull down on the bow and at the same time push the slip forward. At the end of the stroke push the bow upward and pull backward on the slip. This synchroniza-

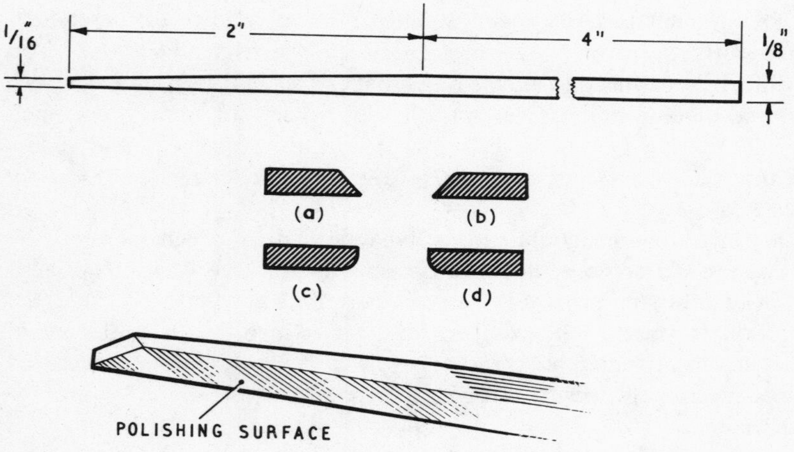

Fig. 32 Polishing slips

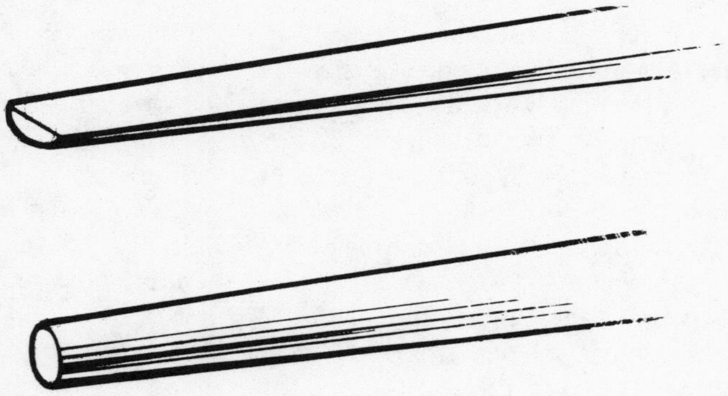

Fig. 33 Polishing slips

tion of movement must be maintained because the slip and the work must always move in opposition.

The co-ordination of movement may feel strange at first but after a very little while it will become easy. Slowly build up speed but do not press heavily on the slip.

Continue this action until all traces of graver marks have been removed, the work will then be ready for polishing.

Remove all traces of oilstone paste from the work by rotating it and holding to it a piece of pith.

Clean the slip thoroughly with a piece of soft rag and then charge it with just a touch of diamantine. Repeat the procedure as before and after a few quick strokes on the bow begin to reduce the pressures on the slip.

Clean the work again with pith to remove the diamantine and the surface should be left with a deep polish.

Square shouldered pivots are best polished with slips that are slightly narrower than the pivot length and, at the same time as the slip moves forward and backward, a small sideways movement is made.

Balance staff pivots need to be polished with curved or round slips, as seen in Fig. 33. When the slip is moved forward it should

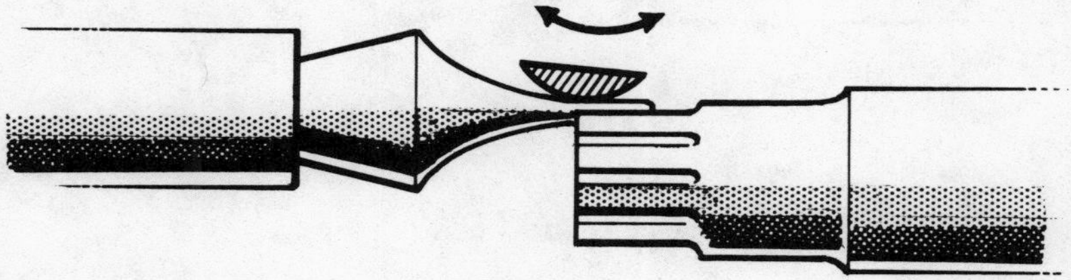

Fig. 34 Polishing balance staff pivot

be rotated slightly in a screwing action and given some sideways movement, as shown in Fig. 34. This ensures that the curve of the conical pivot is maintained. There is the danger however that the end of the pivot will receive less polishing than the top, resulting in the pivot becoming out of shape rather like Fig. 35, which has been drawn exaggerated for clarity. If this should happen, a flat slip with one edge curved will put the matter right.

The procedure for fitting square shouldered pivots and curved conical pivots is the same. Continue polishing with oilstone paste until the pivot end almost enters the jewel hole and then thoroughly clean the work and the slip as previously described.

The same slip is then used with diamantine to produce the high polish that is so necessary to reduce surface friction to a minimum. Continue with the diamantine until the pivot enters the jewel hole with almost no side movement. Again, thoroughly clean the work and the slip.

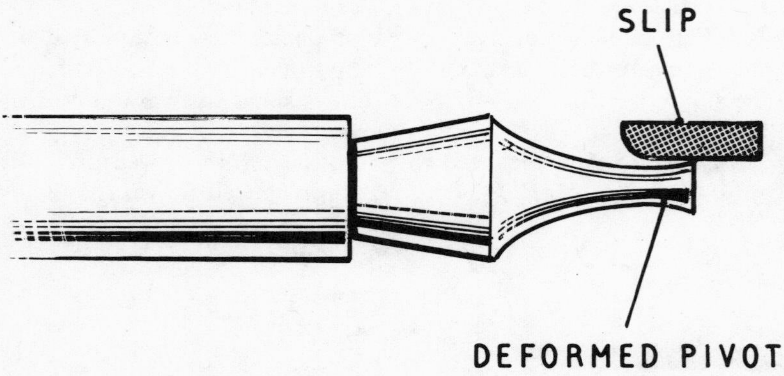

Fig. 35 Polishing a deformed pivot

To impart a finished appearance to pivots it is usual to burnish the ends. In the case of balance staff pivots, it is essential that the ends be burnished to reduce as much as possible surface friction between the pivots and the endstones. For this operation we need a lantern runner, Fig. 36, and of course the burnisher itself.

A piece of steel rod the same diameter as the turns runners is required, and should be about 4 in. long. Turn the ends flat and square in a lathe and into one end drill a hole centrally about ½ in. long and approximately one third the diameter of the runner.

We also need a length of brass rod the same diameter as the steel rod and long enough to be held in a lathe chuck with about ¾ in. protruding. Turn down the brass rod until it is a drive fit in the end of the steel rod.

Reverse the work so that the steel rod is held in the chuck. Cut off the surplus brass rod and turn the face of the brass until the thickness is less than pivot length.

Remove the runner from the chuck and fit it in the turns. Bring the eccentric back runner (male end) up to the brass runner until it makes contact and then rotate the eccentric runner. This will scribe a circle on the brass. Using the circle as a center line drill a number of small holes of different sizes to take pivots.

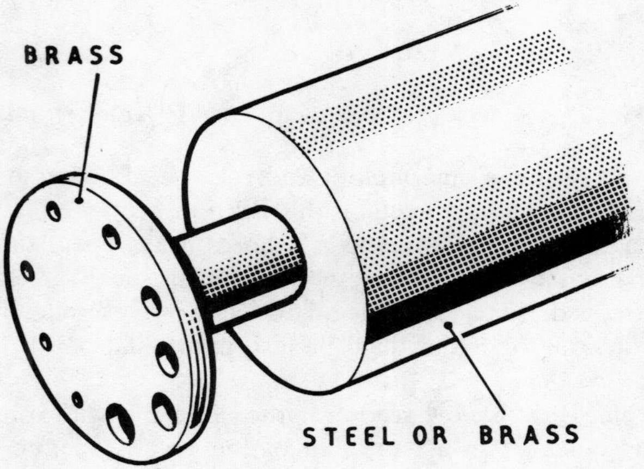

Fig. 36 Lantern runner

Brass is used for the end of the runner to protect the polished pivots from damage. There is no reason why the lantern runner should not be turned completely from one piece of brass rod.

Fig. 37 illustrates a balance staff being supported by a lantern runner in readiness for burnishing the end of the pivot.

Rotate the staff backwards and forwards and give the end of the pivot a few light rubs with an Arkansas stone lightly smeared with thin oil. Keep the stone square to the pivot.

A burnisher for the ends of pivots is a steel blade usually about 2 in. long, ⅛ in. wide and $\frac{1}{16}$ in. thick, set in a long thin round handle.

Good burnishers can be made from worn out files. The cutting teeth are removed by rubbing the file on a piece of coarse emery cloth that has been glued to a flat piece of wood. Smear plenty

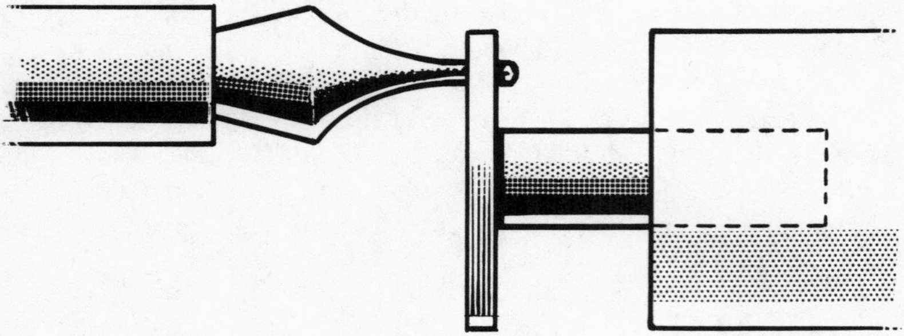

Fig. 37 Balance staff pivot supported in lantern runner

of thin oil over the emery cloth and rub the file over the surface occasionally changing direction; this helps to keep the file flat.

When all the teeth have been rubbed away repeat the process with a smoother grade emery cloth until all deep scratches have been removed. Then continue to use successive grades of emery cloth, each one smoother than the last, until a flat polished surface has been produced.

The burnisher is now ready for use. Smear it with thin oil and place it squarely against the end of the pivot. Operate the bow fairly fast and give the end of the pivot a few light rubs with the burnisher. As the burnisher goes forward turn it very slightly so that the flatness on the end of the pivot is removed, but be careful not to throw up a burr.

The finger-nail test will detect the presence of a burr and the quickest way of removing it is to place the burnisher on top and give the pivot two or three light rubs.

To polish anything flat such as steel springs or brass plates, bridges or cocks, a polishing block and a bolt tool are needed.

A useful size for polishing blocks is 3 in. x 2 in. x ¼ in. The polishing face is prepared by filing flat and is finished with a smooth file. This leaves a fine grain on the metal, which retains the polishing paste. Blocks must be kept clean and when not in use should be wrapped in paper.

Most watch parts that have to be polished are too small to be held in the fingers and so a bolt tool is used for this purpose, Fig. 38.

Fig. 38 Bolt tool

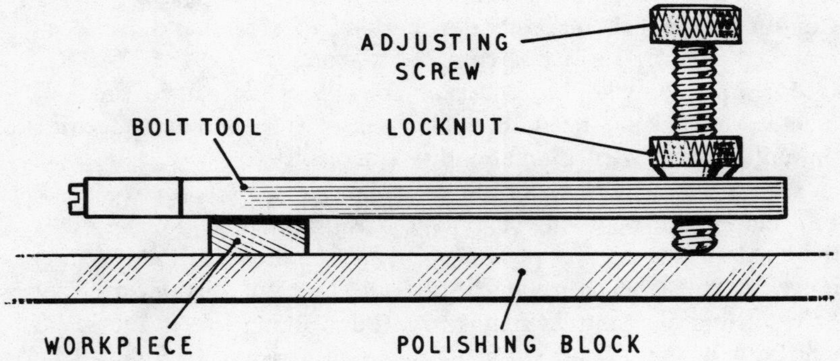

Fig. 39 Bolt tool in use

The watch or clock part that is to be polished is secured to the underside of the bolt tool by shellac, Fig. 39. Warm the tool in a methylated spirit flame and rub a piece of shellac over the surface so that the shellac spreads over an area big enough to take the workpiece. Press the work into the soft shellac and place the tool with the workpiece underneath onto a piece of clear flat glass about ¼ in. thick. Turn the two adjusting screws until the threads protrude underneath by an amount equal to the thickness of the workpiece. Observe through the glass to see if there is a complete face to face contact and readjust the screws accordingly. If the shellac chills before adjustment is complete, it is necessary only to warm the bolt tool again.

When complete contact is made between the work and the glass the bolt tool is ready for use on the polishing block. It is sometimes necessary to drill holes in the under side of the bolt tool to accommodate locating pins, etc., that are part of the workpiece.

Transfer from the mixing block to the polishing block a very small amount of the polishing paste. Replace the storage lid immediately to keep the stock paste free from dust.

Place the bolt tool squarely on the polishing block and move the tool in different directions and in circular motions. Examine the work to see if there is an area not being ground or polished. If necessary make further adjustments on the adjusting screws. Continue to polish until all file marks and scratches have been removed and the surface is completely flat.

When polishing soft or mild steel, or any of the soft or precious metals, there is no need to begin with oilstone powder because the cutting ability of diamantine is sufficient.

This is not so when working with hardened or tempered steel. File marks can be removed only by using oilstone powder; it is the final polish that is produced by diamantine. It follows that the work must be entirely freed of all traces of oilstone powder before using diamantine. If this precaution is not taken, the result will be a high polished surface spoiled by scratches produced by grains of oilstone powder.

Carefully wipe the work and the bolt tool with a piece of soft clean rag and then clean the work with a piece of pith stick. When the pith is dirty, slice off the end and continue with clean pith until all traces of oilstone powder have been removed.

The same process is repeated with diamantine. Begin with some downward finger pressure but concentrate on keeping the bolt tool horizontal with the polishing block. When the surface finish left by the oilstone powder has been removed, the finger pressure can be gradually slackened until the work is little more than stroking the polishing block.

In the course of repair work at the bench, it frequently becomes necessary to join two metal parts together by soldering, particularly in clock repairing. There are two methods in regular use; they are known as soft soldering and silver soldering (hard soldering).

Soft soldering is the most commonly used method. It is quick and convenient, and the parts are readily separated by the application of heat, but it is not as strong as silver soldering.

The method used is known as tinning. The metal faces to be joined are thoroughly degreased by wiping with a cloth moistened with carbon tetrachloride (C.T.C.), or gasoline, and then deoxidized by rubbing with fine emery paper. A final wipe with the solvent will remove any trace of the cleaning agent.

The cleaned surfaces are then coated with resin soldering flux and heat is applied. Do not use spirits of salts (hydrochloric acid) as a flux. It is a powerful corrosive agent and is thoroughly unsuited to this type of fine work.

A small electric soldering iron or a mouth blowpipe will provide the heat source. When the temperature of the metal has reached the melting point of the solder, a length of solder is touched to the work and the solder will immediately melt and flow over the prepared surface.

The work is then removed from the heat and wiped with a clean rag. Surplus solder and flux is wiped away leaving behind a micro thin layer of solder on the workpiece. This completes the tinning operation.

Repeat the process with the other piece to be soldered and then hold the two pieces rigidly together. Apply heat, and when the melting point of the solder has been reached the tinning on the two faces will run into each other. Remove the heat and wait a few seconds until the solder can be seen to flash off. In its molten state solder is shiny, rather like polished chrome. When the temperature drops and the solder returns to solid it takes on the appearance of a matte dull silver color. When this happens the two pieces of metal are bonded together.

Not all metals can be soldered in this way. Aluminum and lead, because of their low melting points, require different techniques. Gold and silver require the application of a solder with the appropriate precious metal base. The two most common metals with which a watch and clock repairer has to work are brass and steel, and both these metals lend themselves well to soft soldering.

The shape of the surfaces to be bonded must present a one hundred per cent contact, e.g., two flat surfaces, or a sliding fit of a rod inside a tube or a tube within a tube.

Hard soldering is a more specialized method. It involves the use of a solder containing silver and is frequently known as silver solder. It has a much higher melting point which precludes the use of a conventional soldering iron. The usual method of applying heat is by a gas blowpipe.

Preparation of the surfaces is the same but the flux used is a mixture of borax powder and water. The two pieces to be joined are clamped together and heat is applied until the borax begins to bubble.

A small piece of solder, which is supplied in thin sheet or strip, is placed on the work and the blowpipe flame continues until the solder turns silvery red and melts. Keep the solder in its molten condition, and with a piece of steel wire flattened at one end, apply a small quantity of borax powder and then, with the end of the wire, encourage the solder to flow the length of the joint. After the soldering the joint must be thoroughly washed to remove all traces of borax and then dried, preferably in boxwood dust.

III

The Turns

THE METHOD OF OPERATING THE TURNS IS RATHER PRIMITIVE and many might think outdated in favor of the modern precision lathe. This may well be so in mass production factories but in small workshops all over the world the turns still take their pride of place. They are slower in operation, but in the hands of an experienced worker they are capable of producing accurate results.

Because they are powered by hand they respond immediately to any change in speed imposed by the operator, and there is therefore less likelihood of damage caused by the cutting tool digging into fast rotating work.

For those with no turning experience the turns will offer the best teaching medium. Even today, most watch manufacturers put their apprentices to work on the turns before giving them experience on the lathe.

It will be seen from Fig. 40 that the turns consist of two centers with adjustable eccentric runners. Both centers are mounted on a common beam but one is riveted at one end whilst the other is free to slide and can be locked in any position. An adjustable sliding T rest is also mounted on the beam between the centers to provide support for the gravers. Power is provided by a hand operated bow, Fig. 41 made from whalebone or cane, and strung

Fig. 40 The turns

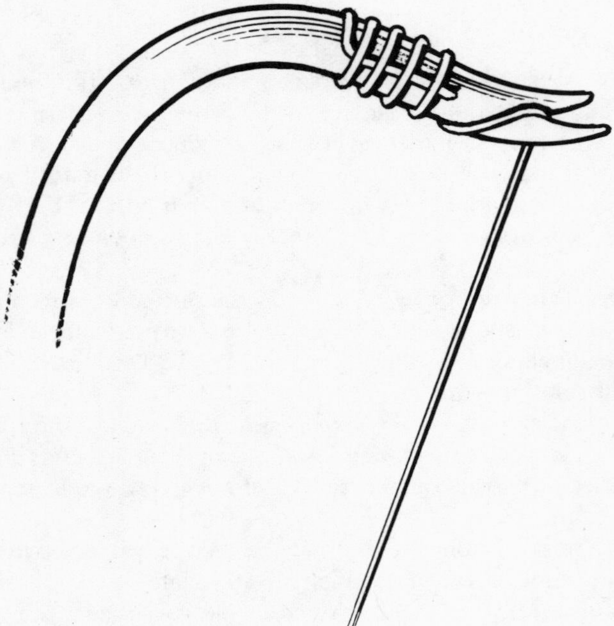

Fig. 41 Bow

with horsehair or nylon thread. When in use the turns are clamped in a vise at the riveted end.

The turns are suitable only for work that can be held between centers or on an arbor. Any work that necessitates the use of a chuck or a face plate can be done only on a lathe but this is no major drawback because most turning jobs can be done between centers.

Work can usually be set up in the turns quicker than on a lathe, and it can be removed from the centers and replaced, with every guarantee that it will continue to run true. This is not so with work that has been removed from the chuck of a lathe and then replaced.

Turning that has been done in the turns usually has a smooth finish whereas lathe turning tends to be ridgy. Polishing in the turns is quicker and easier and is helped by the continual reversal of rotation.

Work to be turned is either fitted with an adjustable screw ferrule, Fig. 42, and then supported between the centers, or if it has a center hole it can be pushed onto a tapered steel arbor, Fig. 43, to which is secured a brass ferrule, and the arbor is held between centers. The adjustable ferrules and the arbors are available in different sizes.

In both methods, motive power is obtainable by winding the bow string one turn round the ferrule and moving the bow backwards and forwards. Fig. 41 shows how to secure the horsehair to the end of the bow. The tension of the string around the ferrule should be enough to overcome the resistance of normal turning but free to slip if the graver digs into the work. This provides a safety factor, often prevents the work from being damaged, and guards against the point of the graver being snapped off.

The cutting tools are hand held and are known as gravers, Fig. 44. They are fashioned from square and diamond section tool steel.

When working on clocks the most useful graver is the square shown at (a). The lozenge graver (b) is used for finishing and cleaning out the corners of inside shoulders. By removing the tip of the point as at (c) the graver is made suitable for rough work. When turning conical pivots the pointed end of a lozenge is

Fig. 42 Ferrule

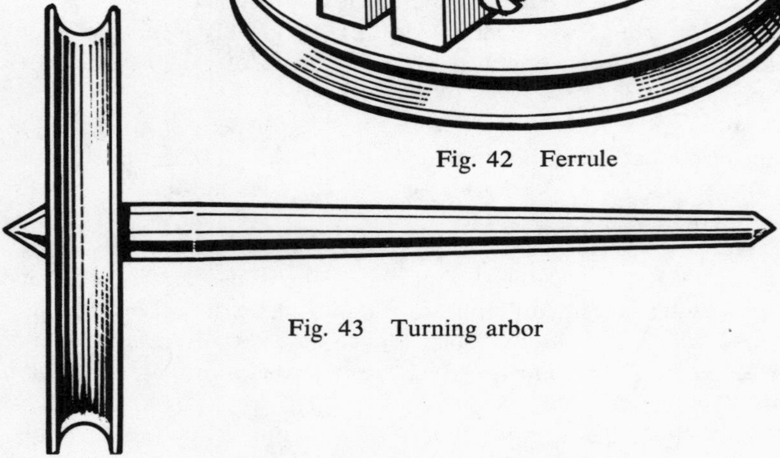

Fig. 43 Turning arbor

rounded off as shown at (d). For deep undercutting a long pointed lozenge (e) is needed. All these cutters are used with the square or diamond face on top. A number three lozenge is a good general purpose size for a watch repairer.

If a graver is used which is too small for the job in hand, it will chatter and produce a rough and out of true finish. On the other hand a graver that is too large is cumbersome and sometimes impossible to use. A little experience will indicate the size that is needed.

Gravers must at all times be kept sharp. Failure to do this will result in spoiled work and loss of time. Sharpen the tools on a smooth oilstone and for very fine work finish the sharpening on an Arkansas oilstone.

Hold the graver so that the square or diamond face is flat

against the stone and then, without changing its attitude, make a series of circular motions covering the surface of the stone. Examine the face of the graver to ensure that it is quite flat, and continue to rub it on the oilstone until burrs can be felt on the edges of the cutting face.

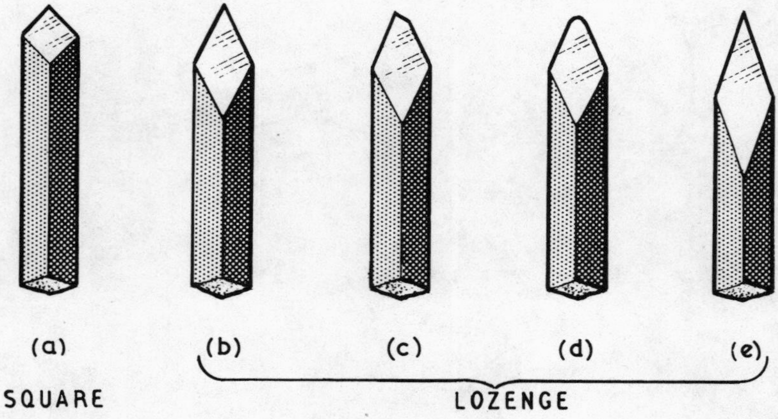

Fig. 44 Gravers

Lay the graver flat on its side on the oilstone and carefully make one stroke to remove the burr. Repeat this on the other side and the graver should be ready for cutting.

If the burr is not completely removed the graver will tend to burnish the work rather than cut it. When this happens to steel, the surface of the work is made harder and the graver tends to push the work away rather than cut it.

When in use the graver is supported by the T rest and is held there and guided by the forefinger. The palm of the hand provides the pressure to hold the cutting edge to the work and therefore a graver needs a small handle.

If the T rest is set too close to the work it will restrict the movement of the graver and will not provide adequate support. If it is too far away the graver will have insufficient support beneath the cutting edge, resulting in rough out of true finishes. The best position for the T rest is about $\frac{3}{32}$ in. from the work.

Cutting is carried out in line with the center of the work as shown in Fig. 45, and the T rest must be adjusted accordingly. If

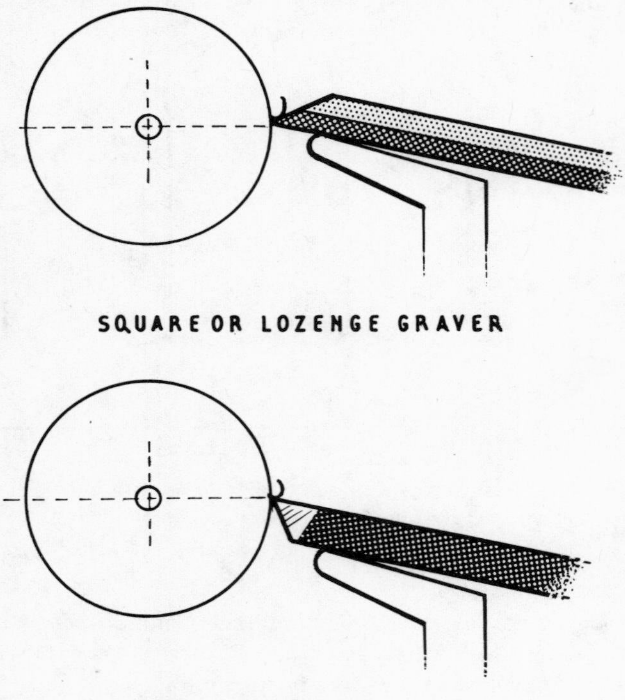

Fig. 45 Position of cutting tools

the point of the graver is set too low it will cause the graver to dig in, and the work will roll onto the upper face of the graver, usually creating irreparable damage and often breaking the point of the graver. If the graver is set too high it will not cut.

Make a start with a short length of brass rod about $\frac{3}{16}$ in. to $\frac{1}{4}$ in. diameter and drill and broach a hole down the center. Fit the brass onto an arbor and mount the assembly between centers. Put a drop of oil on each end of the arbor. If the turns are operated dry the arbors and runners will quickly wear and will run out of true.

Brass needs a faster cutting speed than steel and so a long bow is used. Give the bow string one turn round the ferrule but make sure it is in the correct direction, see Fig. 46. The cuts are made on downstrokes only and the full length of the string is used.

Hold the graver in the right hand with the forefinger extended and the end of the finger on the side of the graver close to the cutting edge. Place the graver and the finger on the T rest and hold the graver as shown in Fig. 66(a). Perhaps the best way of obtaining the correct angle is to place the cutting edge parallel to the work and then pivot on the point to bring the cutting edge away from the work two or three degrees. Then when the graver is moved along the work, as indicated by the arrow, it will be the edge of the tool immediately behind the point that will be cutting.

Take the bow in the left hand and position it for a downward stroke. Remember it is only on the downward stroke that cutting can take place. On the upward stroke the work will be rotating in the wrong direction and the graver must be eased away from the work. This is done by rolling or pivoting on the end of the finger without altering its position on the T rest. The amount of movement needed is so slight that it can be felt better than seen.

If the work is not round or if it is eccentric than the graver must be held firmly to the T rest, at such a distance from the work that cutting can take place only on the high spots. As successive cuts are made, the shavings will become longer and the work will come nearest to the round until, when the work is quite round and true, cutting will become continuous.

Pull down on the bow and make your first cut. Slight variation to the angle of the graver during cutting will indicate the best position. You will be able to feel when the metal is coming away clean and leaving a smooth finish. If there is any chattering or digging in, stop rotation and investigate. The fault will be a blunt graver or one that has been incorrectly applied, or both.

When the diameter of the workpiece has been reduced, the resultant shoulder will need squaring. Move the graver round

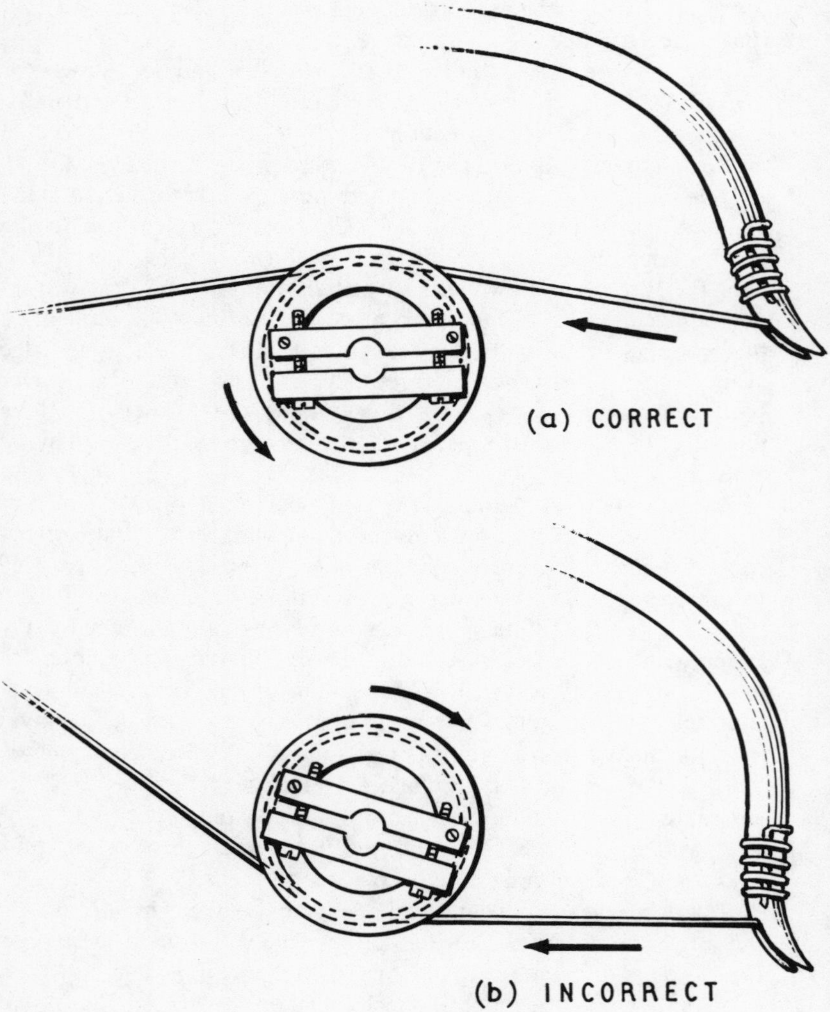

Fig. 46 Stringing a ferrule

into the position shown in Fig. 66(b), begin cutting at the root and then draw the graver towards yourself. The cutting action is the same as before but the opposite edge of the graver is now being used.

Turning a Balance Staff

A common cause of watch failure is a broken balance staff pivot and this seems an appropriate subject for an exercise. We will require an old balance assembly so that a new staff can be made to the exact dimensions of the original.

When a watch needs a replacement staff the quickest and most economic repair is to fit a factory replacement, a typical example of which is illustrated in Fig. 47, but these are not always readily available. The next best thing is to obtain from your material dealer an oversize blue steel staff, Fig. 48. These rough staffs are made of silver steel which have been hardened and then tempered to a dark blue color.

Failing these two sources of supply, the only other course of action is to turn one from a length of silver steel wire. The wire can be either soft steel or blue steel. Soft wire is used only for producing a rough staff at which stage it is hardened and tempered. If blue wire is used, the hardening and tempering will already have been done.

It is convenient to have at our disposal the original staff from which measurements can be taken, and we must therefore begin by removing the old staff from the balance.

Before any disassembling is carried out it would be advisable to make a sketch showing the relative positions of the balance spring collet, the bar of the balance, and the roller. If, when the new staff has been made, re-assembly is done according to your drawing, it might save a lot of time.

Begin by taking off the balance spring complete with split collet. The correct method is by inserting a wedge into the slot in the collet and opening it enough to lift it from the staff. It is not good practice to insert the blade of a knife between the collet and the shoulder on the staff and lever off the collet. This method can disfigure the collet and frequently results in damage to the collet and the balance spring.

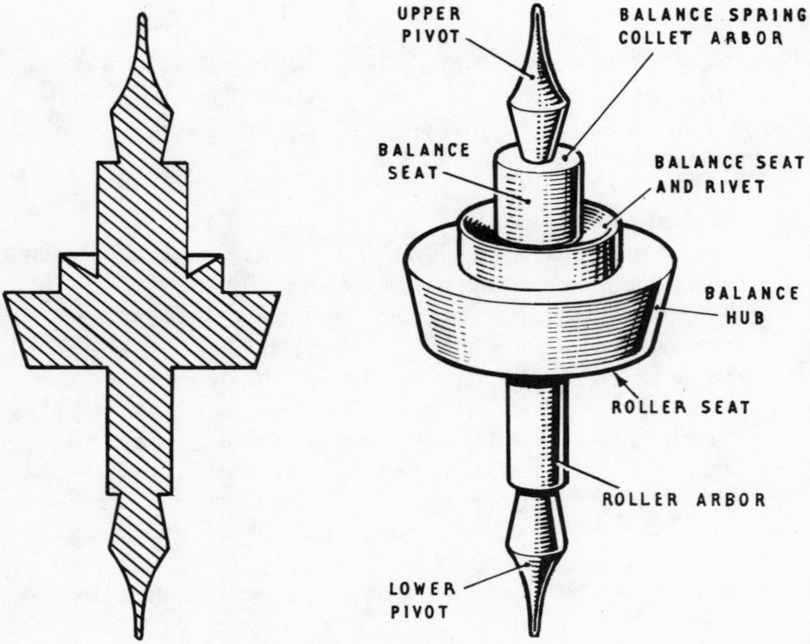

Fig. 47 Balance staff

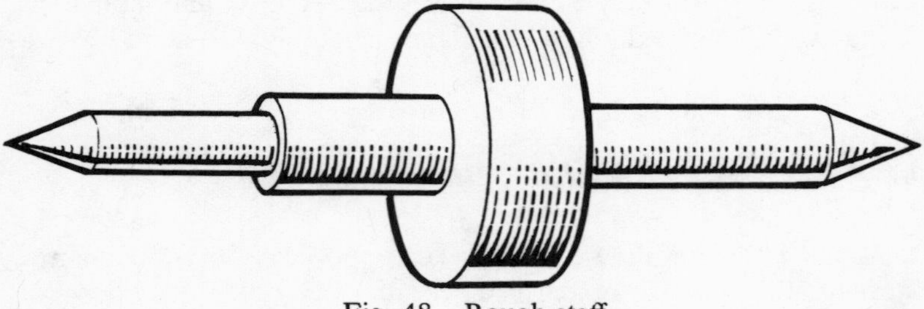

Fig. 48 Rough staff

It is safer to remove the collet by using a wedge. Take a short length of stainless steel wire between 0.010 in. and 0.025 in. diameter, according to the size of the collet, and file two flats at one end like a screwdriver but with a much longer taper. The thickness at the base of the taper should be a little less than the width of the slot to permit ease of entry. Insert the other end of the wedge into a short length of pegwood stick for a handle.

Do not push the wedge to the bottom of the slot because it

may open the collet too far causing metal fatigue and rendering the collet unfit for further use.

Insert the wedge just enough so that when it is given a slight upward twist it will lift the collet and balance spring from the staff. Hold the collet to the bench with tweezers and the wedge can then be pulled out.

If, as sometimes happens, you meet with a situation where the slot in the collet is too wide for a wedge to open, then there is no alternative but to pry the collet from the staff.

Take a piece of silver steel wire and file it as for a wedge, with a long taper; but this time shape it to a knife edge.

Hold the balance in the finger-tips and very carefully insert the lever between the collet and the staff. Begin to pry the collet and then change the position of the lever so as to keep the collet as square as possible with the staff. The last movement should be done close to the bench top over a piece of clean white paper.

Next, hold the staff around the balance spring arbor in a pin vise, and hold the roller in a second pin vise. Very carefully pull and twist and the roller will come away from the staff.

You may not have a pin vise large enough to hold the roller but if you have there is no guarantee it will be suitable for the next roller removal job you do. Take a look at some of the tool suppliers' catalogs and you will see a range of roller removers, some of which are inexpensive. They are safer to use than a pair of pin vises and can be time-savers if you are likely to be involved in much of this type of work.

All that now remains is to remove the staff from the balance. Set up the staff between the runners as shown in Fig. 49, and then, with a pointed lozenge graver, undercut the rivet until only a thin shell of the original riveting metal remains.

Use either a riveting stake, Fig. 50, or a staking set, Fig. 51, and select a hole slightly larger than the diameter of the balance hub of the staff. Place the staff in the hole with the balance uppermost. Select a punch with a hole that will just fit over the top pivot. Give the punch a light tap with a hammer and the staff will be driven out of the balance.

Now that the staff is free it can be used as a pattern from which to take our measurements. Because we are assuming that the staff

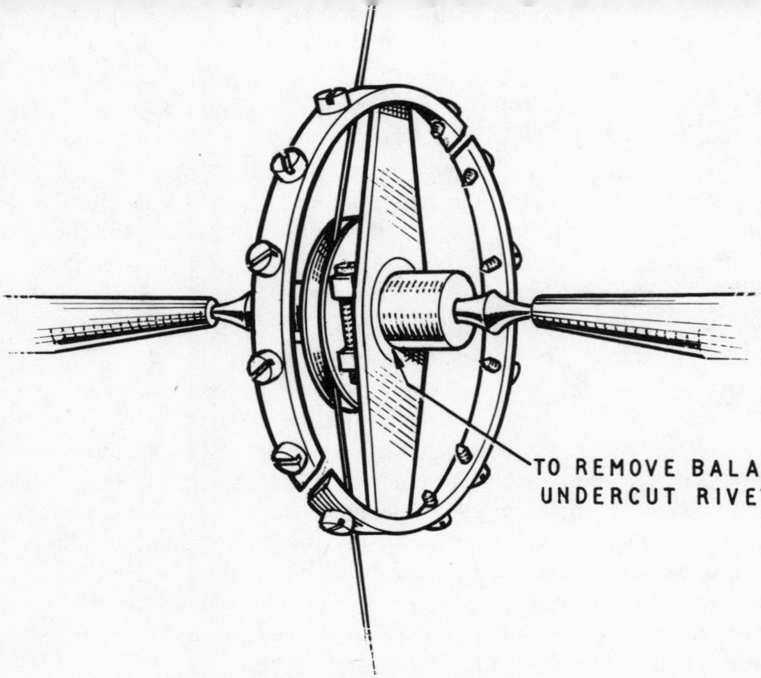

Fig. 49 Removing balance from staff

has a broken pivot, we cannot measure the staff to obtain its overall length. This we have to obtain from the movement. In order to do this, we need a home made screw gauge similar to the one illustrated in Fig. 52.

Slacken both locking screws and screw the short thread into the frame as far as it will go. Remove the end stones from the jewel holes and place the end of the short thread over the lower jewel hole. Screw in the long thread until it touches the upper jewel hole. Unscrew the short thread to release the gauge from the movement and then return the short thread to its former position. The distance between the ends of the two threads is the overall length of the staff.

Alternatively the measurement can be taken with either a degree gauge, Fig. 20, or a vernier sliding gauge, Fig. 21.

Let us assume we have a blue staff which is slightly oversize in all dimensions. Fit up the staff in the turns as shown in Fig. 53, adjust the T rest and, with a pointed diamond shape graver, turn

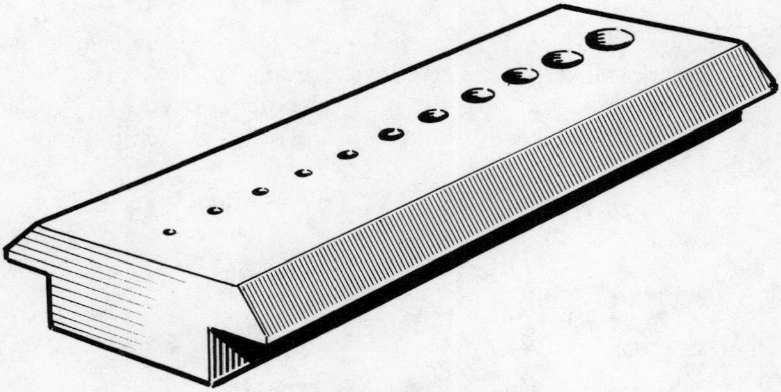

Fig. 50 Riveting stake

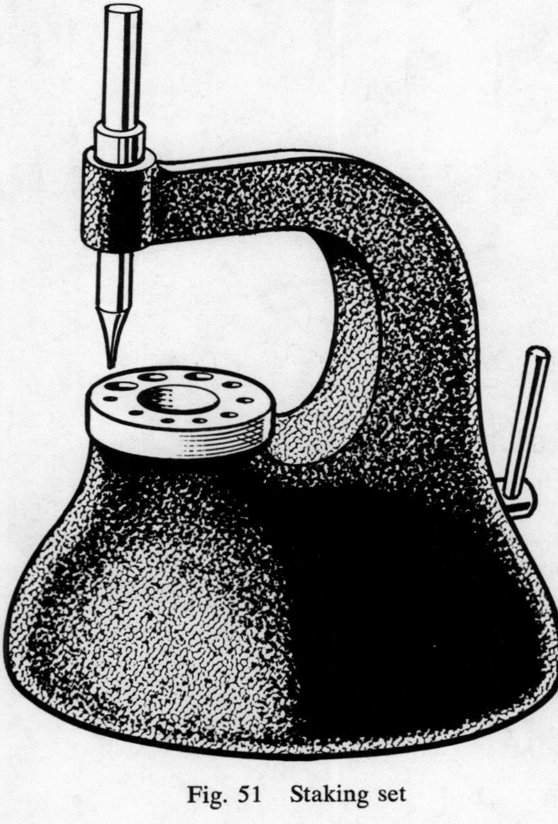

Fig. 51 Staking set

the back slope to an angle of about 45 degrees to the center line of the work. Make the last few cuts very fine so that a good finish is obtained. This will help the next operation, which is to polish the back slope. Details of polishing in the turns is given in Chapter II.

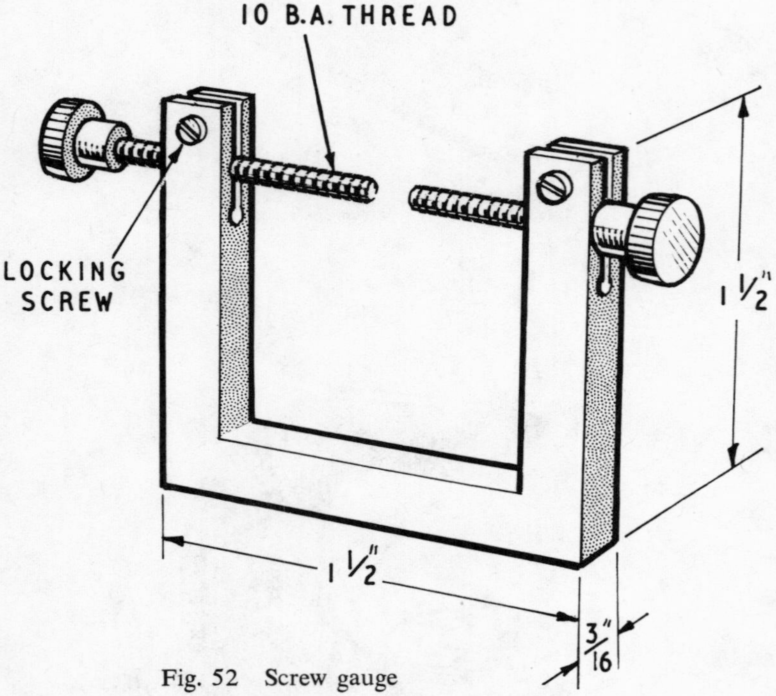

Fig. 52 Screw gauge

Remove the work from the runners, take off the ferrule and fit it on the roller arbor, and then refit the work in the turns. The balance arbor and seat are now turned down.

Remember we are using the original staff as a pattern from which we can take our measurments. To measure the various diameters of the original staff we will need either a micrometer, Fig. 21, or a pair of pinion calipers, Fig. 54. Instructions on how to read a micrometer are given in Chapter I, so it is enough to say here that

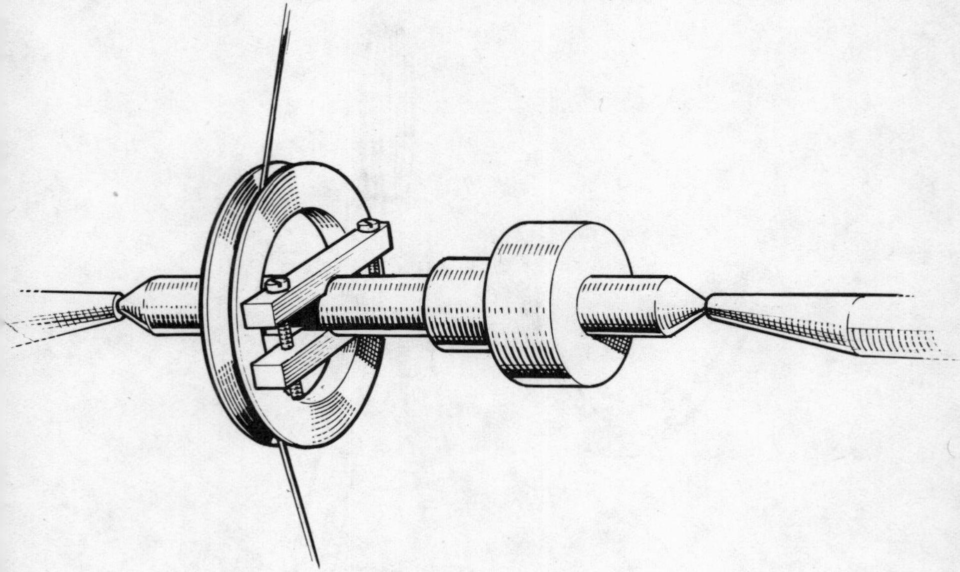

Fig. 53   Blue staff ready for turning

whereas a micrometer records diameter in units of measurement, the calipers act only as a go and not-go gauge.

The diameter of the arbor should be reduced to within about 0.002 in. of the original staff, and at this stage the balance should almost fit. Then a very slight taper of approximately 0.0005 in. along its length should be given to the arbor. If calipers are being used, the new arbor should be turned down until the calipers almost pass over the arbor. Care must be exercised when trying the balance for fit because if the balance is too tight it can easily be damaged by excessive finger pressure. Turn the arbor until the tapered end enters the hole in the balance and the rest of the fit is just enough to prevent the arbor from falling out.

It is essential that the shoulder of the balance seat is square and sharp, and that no lump or ridge is left in the inside corner. If a ridge was left when the balance was riveted to the staff, the balance would take up a position on the staff that was out of square, and

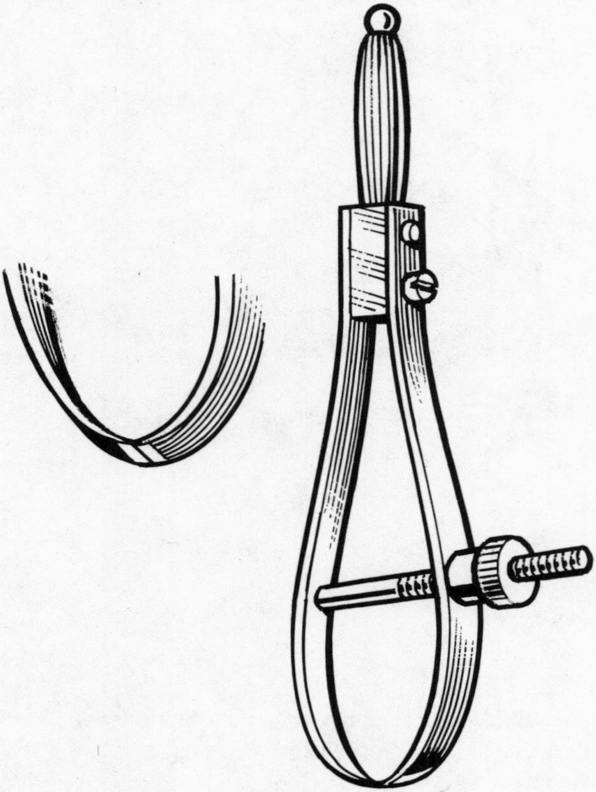

Fig. 54 Pinion calipers

it may even rock. To ensure that this does not happen it is usual to slightly undercut the inside corner.

When the balance is in position on its seat, rotate the work and mark it with the point of the graver to indicate where the shoulder of the collet arbor is to be formed. The balance arbor must be allowed to protrude slightly beyond the balance to provide the metal necessary for the rivet, and it is here that the graver mark is made.

Turn down the collet arbor finishing with a very slight taper as was done with the balance arbor. Maintain this taper until the collet can be pushed with the fingers along the arbor to a distance of approximately twice its thickness from the shoulder. Then undercut the face of the balance arbor to produce the rivet as shown in Fig. 47. The collet arbor is then polished until the collet can be pushed up to its seat with a sliding fit.

Now reverse the work in the turns with the ferrule fitted to the collet arbor, and turn the roller arbor. Finish with a very slight taper until the roller can be pushed along the arbor to a distance of approximately its own thickness from its seat. If the roller is forced it will most probably become damaged. Make sure the arbor shoulder is turned square and the inside corner is clean, allowing the roller to seat correctly. Polish the arbor and keep trying the roller with tweezers until a nicè friction fit is obtained.

To turn the top pivot the ferrule is fitted to the outer end of the roller arbor and the work is reversed. Using the original staff as a pattern mark the work with a pointed graver to show where the pivot is to finish, Fig. 55. The excess metal is cut off with a pointed diamond graver as shown in Fig. 56. When working in the turns or on a lathe, to reduce the length of a piece of metal in this way is known as parting.

Because the work is supported only at the ends it follows that as the cut deepens the amount of support becomes progressively less until, when the cut is on the point of completion, the work collapses. To prevent this from happening, deep cuts must be avoided, and at the first signs of the work losing its stability between the runners further turning must be stopped. The excess metal can then be broken off with the fingers but a small flat will be left at the end of the cone which must be removed before we turn the pivot.

Take the female runner from the right of the turns and fit a home-made V bed runner in its place as shown in Fig. 57. A piece of silver steel rod will be required the same diameter as the original runners and about 4 in. long. File it to shape and polish the bearing surface of the supporting V bed to prevent undue marking of the work. Harden and temper the end and clean with an emery stick.

Fit the work up in the turns with the V bed runner on the right

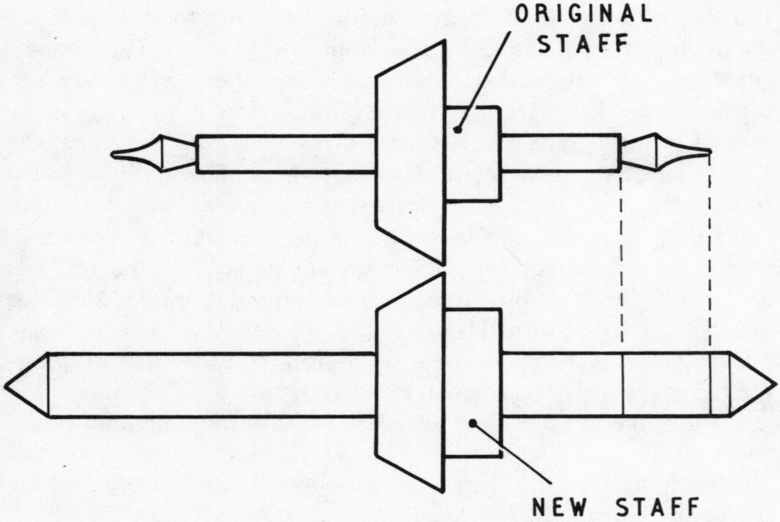

Fig. 55 Marking new staff

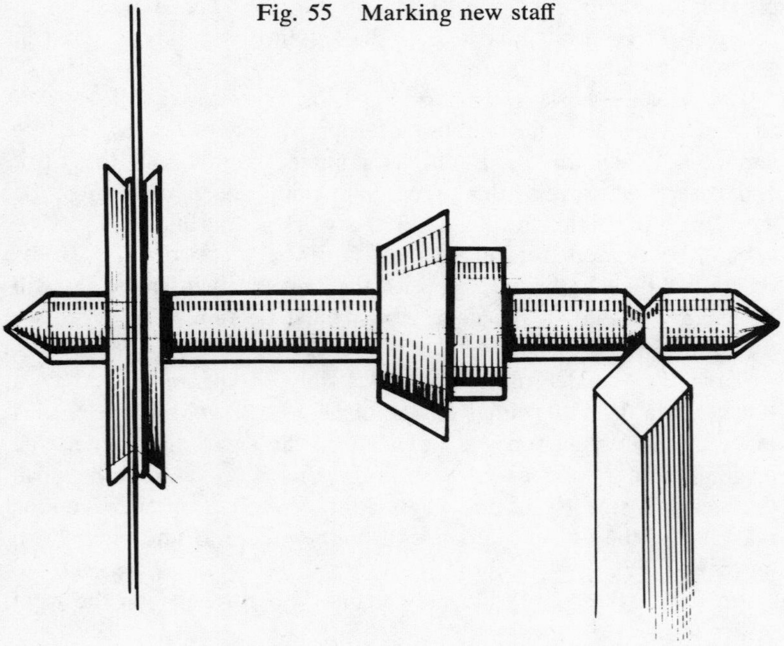
Fig. 56 Cutting new staff to length

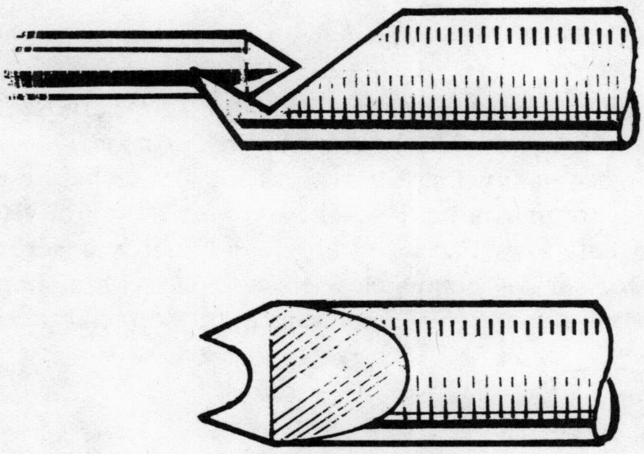

Fig. 57 V Bed runner

and with the ferrule still in position on the roller arbor on the left. Spin the work and, with an Arkansas slip, stone the conical end of the staff to a sharp point.

We are now ready to turn the top pivot. Remove the V bed runner and fit up the work between the original female runners. Here again we can use the old staff to indicate where on the collet arbor we must begin the top pivot. Mark as before, using a pointed graver and cut the back slope of the pivot as shown in Fig. 58. Now turn the cone of the pivot so that the curved

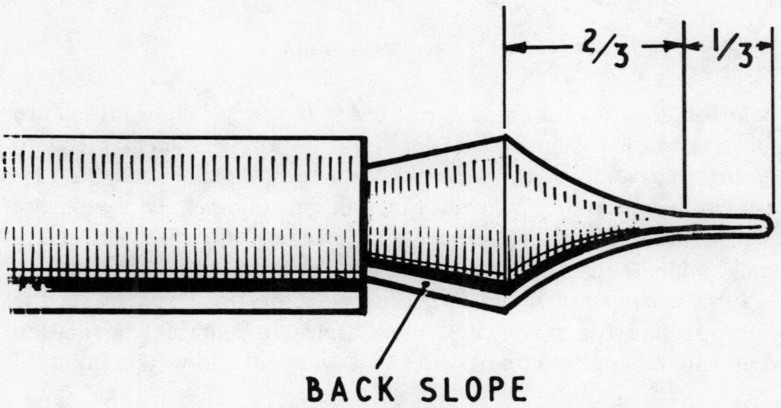

Fig. 58 Shape of finished balance staff pivot

section takes up approximately ⅔ of the length and the remaining length is parallel.

The original runners supplied with the turns are designed for general work. They are too big to allow fine pivots to be turned and so a pivot runner must be made for this purpose. Once it has been made it should be kept for future occasions.

Take a piece of silver steel rod the same size as before and turn the ends square in a lathe. Grip the rod vertically in a vise that is fitted with soft jaws, brass or fibre, and punch a series of indentations of varying depths close to the edge with a sharp center punch as shown in Fig. 59. The bruising of the metal around each

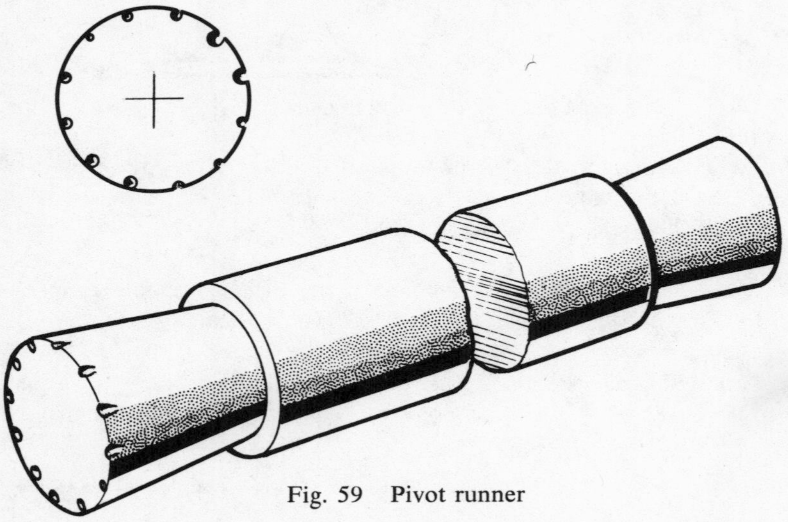

Fig. 59 Pivot runner

hole forced up by the punch is removed with a slip stone. Turn the ends with a slight back slope and remove sufficient metal to ensure that some of the indentations almost break through. This operation will have to be carried out on a lathe. The ends are then hardened and tempered and cleaned up with an emery stick, and the runner is ready for use.

Fit the new staff up in the turns with the pivot runner on the right. Rotate the pivot runner and select a suitable indentation that will not only support the pivot but will allow the point of the graver to reach the pivot end. Continue to turn the parallel section of the pivot until it almost enters the jewel hole in the balance cock. Final fitting is done during polishing.

We now have a slender pivot on the end of our new staff and adequate support underneath must be provided before any polishing can be attempted. Here again a special runner is needed.

A piece of silver steel rod, as before, is turned in a lathe to bring the end faces square, and then the end faces are marked in the center. The rod is then held in a vise and, with a pair of spring dividers, a circle is scribed on each end. This circle is to be the center line for a series of holes. Fig. 60 illustrates what is

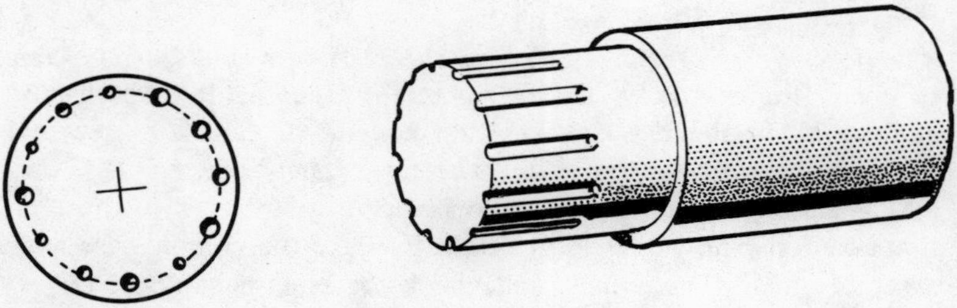

Fig. 60 Polishing runner

required. Each hole is a different diameter and should be drilled to a depth of about twice the length of a pivot.

When the holes have been drilled, each end is turned in a lathe until the diameter has been reduced to the scribed line. The holes will then have become semi-circular grooves. Remove any burrs with the emery stick and then harden and temper each end. Final cleaning with the emery stick completes the runner, and if a careful choice of drill sizes has been made it will prove to be a most useful accessory to your turns.

Fit the staff up in the turns with the grooved runner on the right. Select a groove that has a depth of about half the diameter of the pivot and rotate the runner to bring the selected groove on top. Make sure that the pivot is supported only beneath the parallel section. The curved shoulder must be clear of the grooved runner; otherwise, when polishing takes place, the downward pressure will bend the pivot and the curved shoulder will be ridged by the end of the runner.

The staff is now ready for the top pivot to be polished and burnished. Instructions are given in Chapter II on how to make and use the necessary tools, and how to apply the polishing materials.

When the top pivot is finished, the staff is reversed in the turns and the screw ferrule is fitted to the outer end of the balance spring collet arbor. Place the original staff against the work and mark the position for the bottom pivot. The procedure for shaping the bottom pivot and polishing and burnishing is the same as for the top pivot. For easy reference the sequence of turning operations is illustrated in Fig. 61.

Before we rivet the balance to the staff it would be advisable to check the accuracy of our work. Replace the jewel hole end stones and fit the new staff to the movement. Check to see that endshake is present but not excessive. Then remove the staff and fit the balance, the split collet and the roller to it, in the relative positions shown in your sketch. Replace the complete balance assembly in the movement and check for alignment and freedom of movement. The balance spring stud can now be refitted to the balance cock, and the movement set in motion. Any adjustments that are necessary are made at this stage. When the balance is functioning correctly the assembly is removed from the movement and the split collet and the roller are removed from the staff. The balance can now be riveted.

Place the staff on a riveting stake with the roller arbor in the smallest hole possible, Fig. 62. Select a bevelled hollow punch that fits snugly over the balance spring collet arbor and give the punch a few light taps with a hammer, enough to spread the metal. If you have a flat bottom punch of the same size this can be used to finish the rivet. The balance can now be re-assembled and put back in the movement.

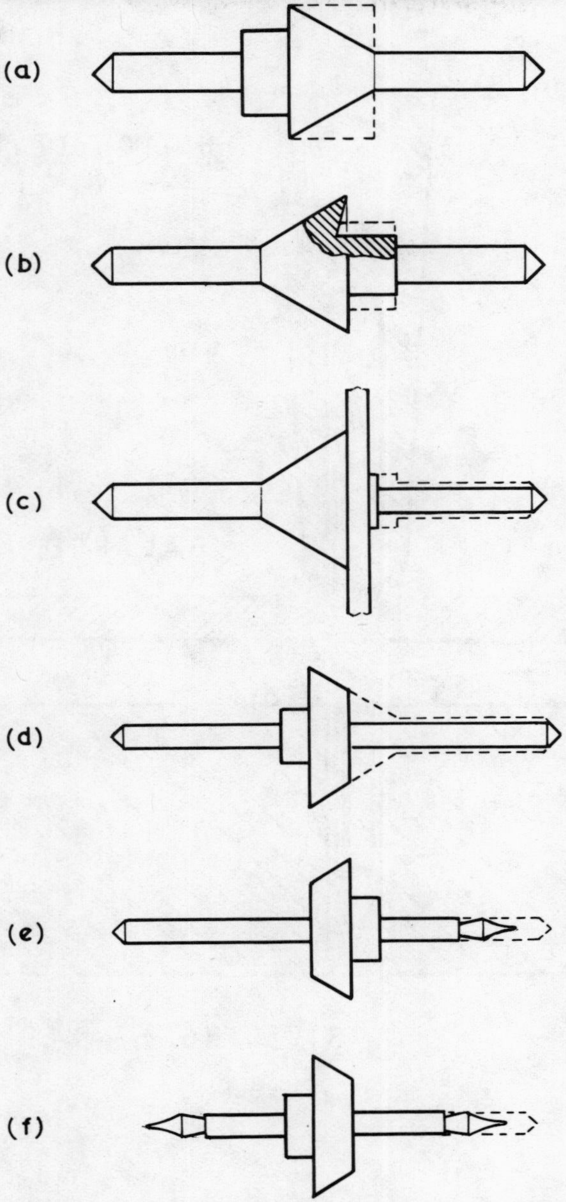

Fig. 61 Sequence of turning balance staff

(a) Turn and polish back slope.
(b) Reverse. Turn balance seat.
(c) Turn and polish arbor for split collet.
(d) Reverse. Turn and polish roller arbor.
(e) Reverse. Cut staff to length, and cut and polish top pivot.
(f) Reverse. Cut staff to length, and cut and polish bottom pivot.

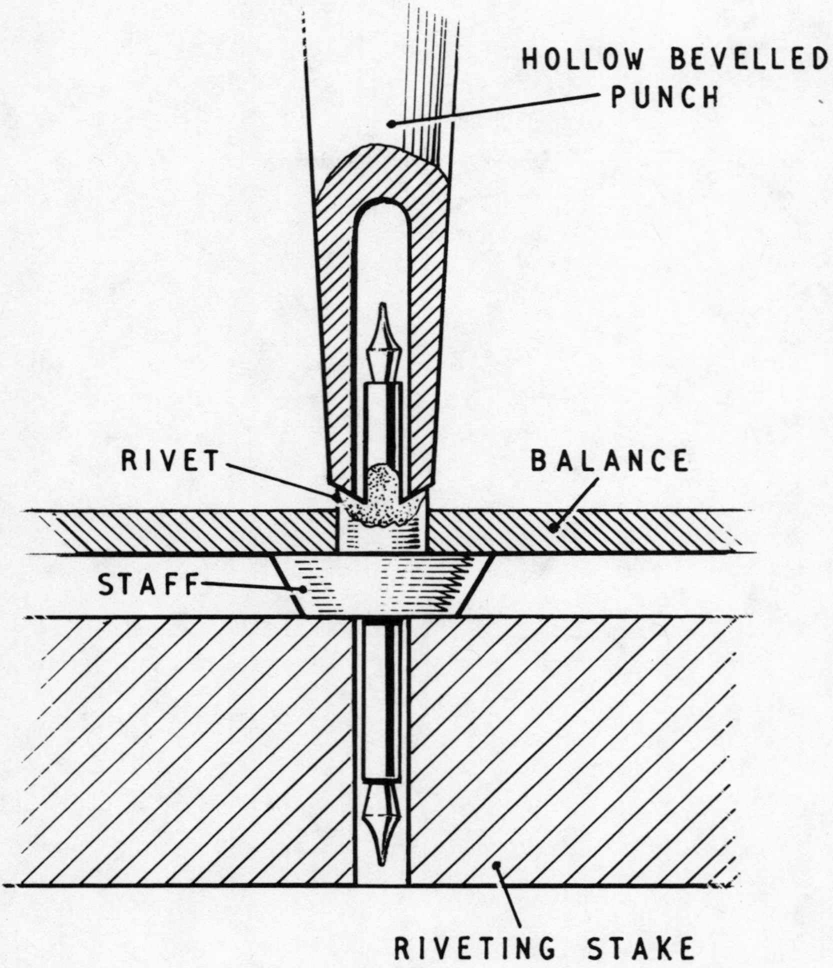

Fig. 62 Riveting staff to balance

IV

The Lathe

A WATCHMAKERS' LATHE IS A GREAT ASSET TO A REPAIRER. LATHE work is probably the one phase of an apprentice's training to which he looks forward more than any other. Once mastery of the machine has been achieved and practice has produced the required skill, one can dispense with many of the services of outworkers and become more independent and self-supporting.

It will take days, sometimes weeks, for a repair or replacement to be returned to you by mail. With your own equipment many such jobs can be accomplished in a matter of minutes. The cost of the finished job is less and there is a considerable reduction in the time taken to complete the work, both of which have to be considered when working for profit.

Modern watchmakers' lathes, with their range of accessories, can provide the man at the bench with a comprehensive machine shop capable of many jobs other than turning.

Sometimes it is better to buy a secondhand machine that, although less sophisticated, is nevertheless capable of most operations. Certainly there is a considerable saving in financial outlay.

Most of the older types of lathes are American or German. The Germans favored the use of a stout bar as the bed around which the parts were mounted. The components were kept in a vertical position and prevented from turning by sliding in a square channel cut along the length of the bar, Fig. 63. The lathe could be mounted in its own pedestal, which was screwed to the bench top, or it could be clamped in a vise.

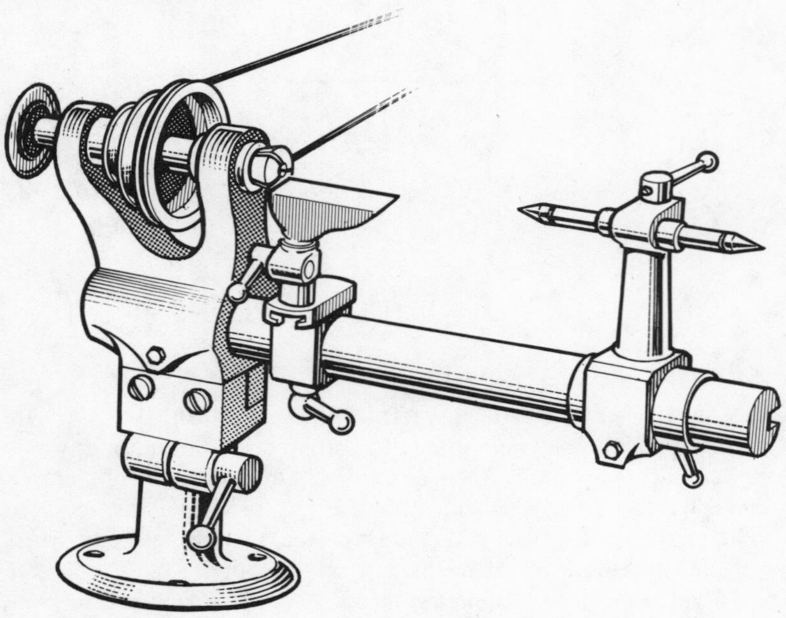

Fig. 63 The lathe

The American lathe was larger and heavier and was made with a broad flat bed. The centers were higher than those of the German design and the lathe could carry a larger diameter face plate making it suitable for almost any type of watch work.

Motive power is supplied by means of a hand wheel mounted on the underside of the bench top, a foot operated treadle wheel fitted to the floor, or a small electric motor mounted on top of the bench, Fig. 64.

The hand wheel method is still very much in evidence in older workshops. The operating speed is slower than that of the other two methods and consequently greater control can be exercised. It is ideal for a beginner.

The treadle wheel calls for some practice before the action of the foot becomes automatic, but once this has been mastered the operator has the same degree of control as with the hand wheel, plus the advantage that both hands are free.

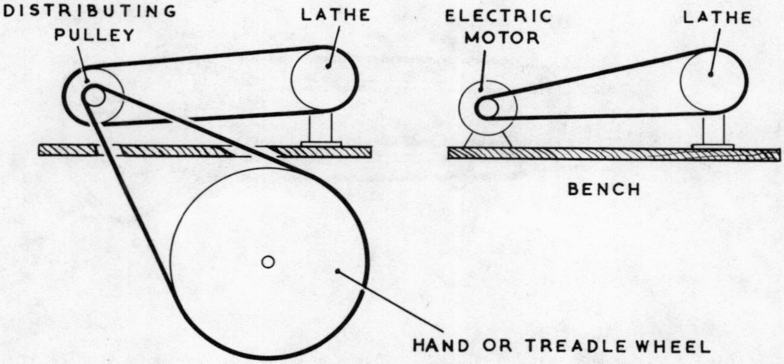

Fig. 64 Motive power

The electrically powered lathe requires a different handling technique and greater care is needed during use. It is fast, and removes metal during cutting and polishing more rapidly than does a hand or foot operated machine. Because of the speed the slightest jerk or dig by the cutting tool invariably causes irreparable damage to the work and frequently breaks the point of the cutting tool.

Of the three methods of supplying motive power there is no doubt that the electrically driven lathe is the most efficient.

Old though a machine may be, if it has been well cared for it should be capable of producing accurate work. Any slight wear caused by normal usage can usually be taken up by adjustment.

When choosing a secondhand machine the general condition will often suggest the kind of treatment the machine has had. Look for burrs or bruising of the metal that might indicate the lathe has been dropped.

Fasten the lathe to a bench. Inspect the jaws of split chucks, Fig. 65, sometimes called collets or wire chucks, for damage. Make sure the threads are deep cut and in a good condition, and that the tops of the threads are not worn away. Test each chuck in the draw-in spindle. Insert the chuck into the headstock with the keyway in the chuck engaging with the key in the headstock, and finger tighten the draw-in spindle. The chuck should be pulled in to the conical seat and the jaws should close evenly.

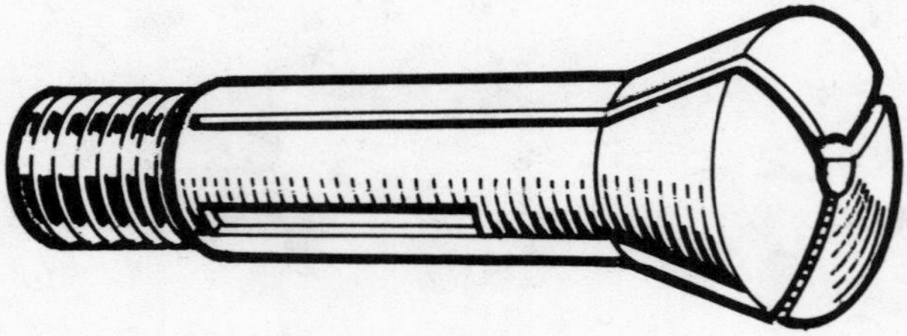

Fig. 65 Split chuck

The headstock and tailstock must be in alignment. Insert the smallest size split chuck you have into the headstock, and screw up the draw-in spindle. Fit a male center runner into the tailstock and slide it up to the headstock. Note the position of the male center in relation to the center hole of the chuck; they should coincide. If they do not, it is an indication that the headstock or tailstock are not sitting squarely on the bed, or they or the bed may be twisted. The headstock or tailstock may at some time have received a blow causing permanent distortion. In any case there is nothing you can do about it and you would be advised to look for another lathe.

Check the headstock for end play, as there should be no perceptible movement. If there is, this is no reason for rejecting the machine providing there is adequate adjustment available on the bearings.

Turning can be carried out between centers using gravers the same as with turns, but it is more usual to hold the work in a chuck and turn either with a graver, Fig. 66, or with cutters designed for use in a slide rest, Fig. 67. These cutters are clamped in position which gives them greater rigidity, and because they

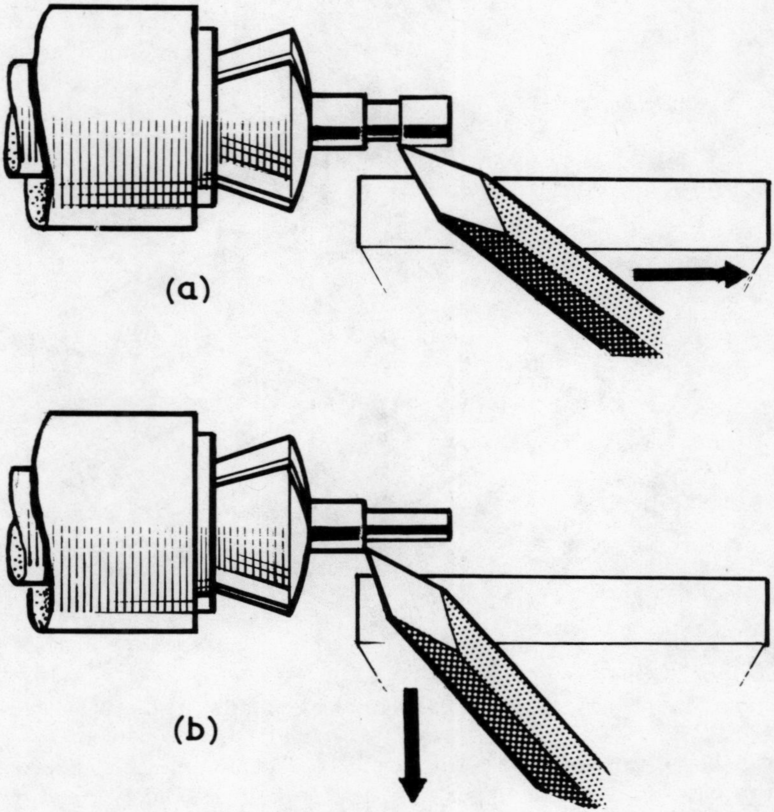

Fig. 66 Turning with a graver

can move only in a straight line, the arrangement produces accuracy with less effort.

The slide rest consists of a steel base which can be slid along the bed and clamped in any position. Mounted one above the other on the base are two slides, each capable of being moved in dovetails by means of handle operated lead screws.

The lower slide moves back and forth, and crosses the bed at right angles. The upper slide moves parallel to the lathe bed, but

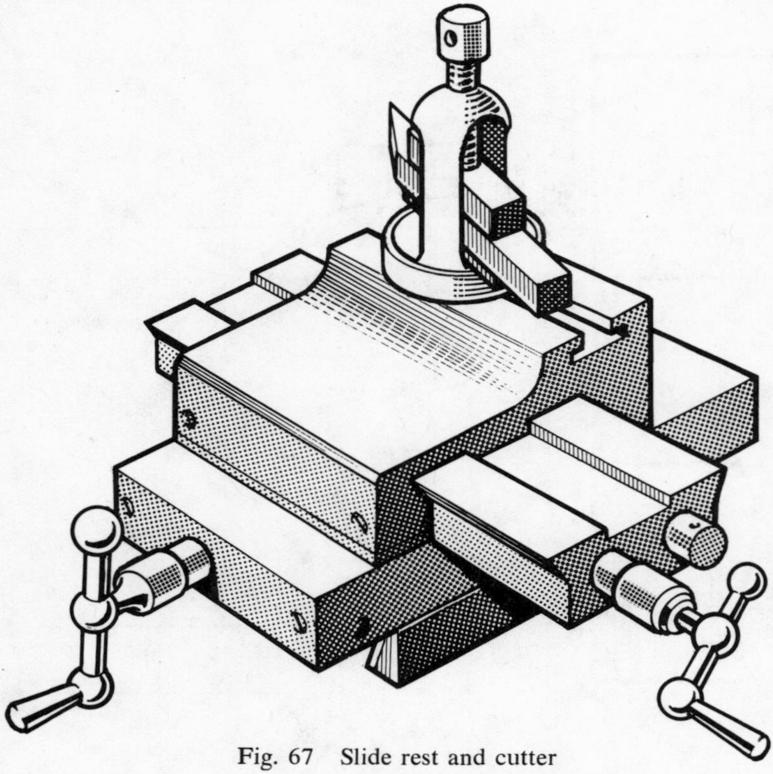

Fig. 67 Slide rest and cutter

can be adjusted to move within ninety degrees on either side of this basic setting. The upper slide carries the tool holder.

Secure the slide rest and try to move the slides by hand without turning the lead screws. There most be no movement. If there is, it will be impossible to produce accurate machining.

Wear in the dovetails can be taken up by screwing in the adjusting bar screws, Fig. 68. Wear in the lead screws and the lead screw guides can be eliminated only by replacement parts.

It is advisable to disassemble a secondhand machine and thoroughly wash all bearings and oilways in kerosene. After this, it should be reassembled and the headstock bearings adjusted to give complete freedom without end movement.

The drive is taken by cat gut or nylon gut, both of which are

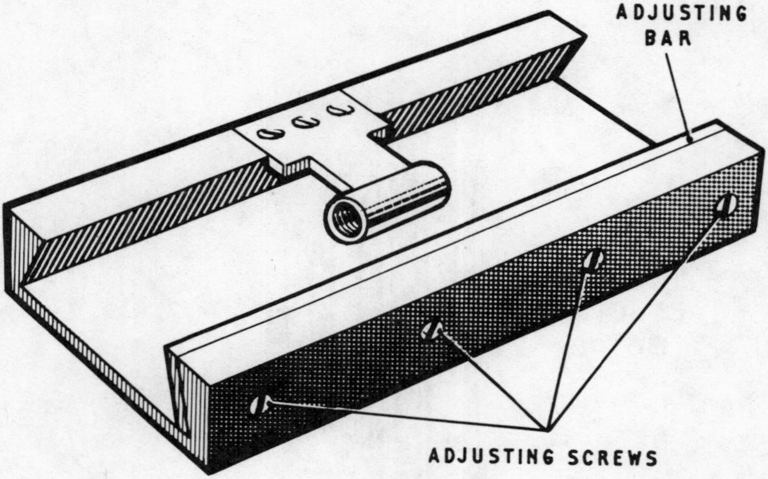

Fig. 68 Dovetail slide adjustment

obtainable from clock material dealers. The arrangement for belting is shown in Fig. 69.

Tools intended for use in slide rests vary considerably in shape as a glance at an illustrated tool catalog will show. However, the cutters shown in Fig. 70 will meet the requirements for most jobs.

As with gravers the cutting edges must be sharp. The height of the cutting edge should correspond to the center line of the work, and can be adjusted by the introduction or removal of metal strips or shims beneath the tool.

To prevent chattering of the cutter and resultant rough finish, the tool must be firmly secured by the clamping screw and the amount of overhang must be kept to a minimum.

When drive belts and pulleys are used, it is common to provide means of altering the machine speed without altering the speed of the driving force. This is done by varying the diameters of the pulleys and using alternative pairs, Fig. 69. A small diameter driving a large diameter will produce a slower speed than when the situation is reversed.

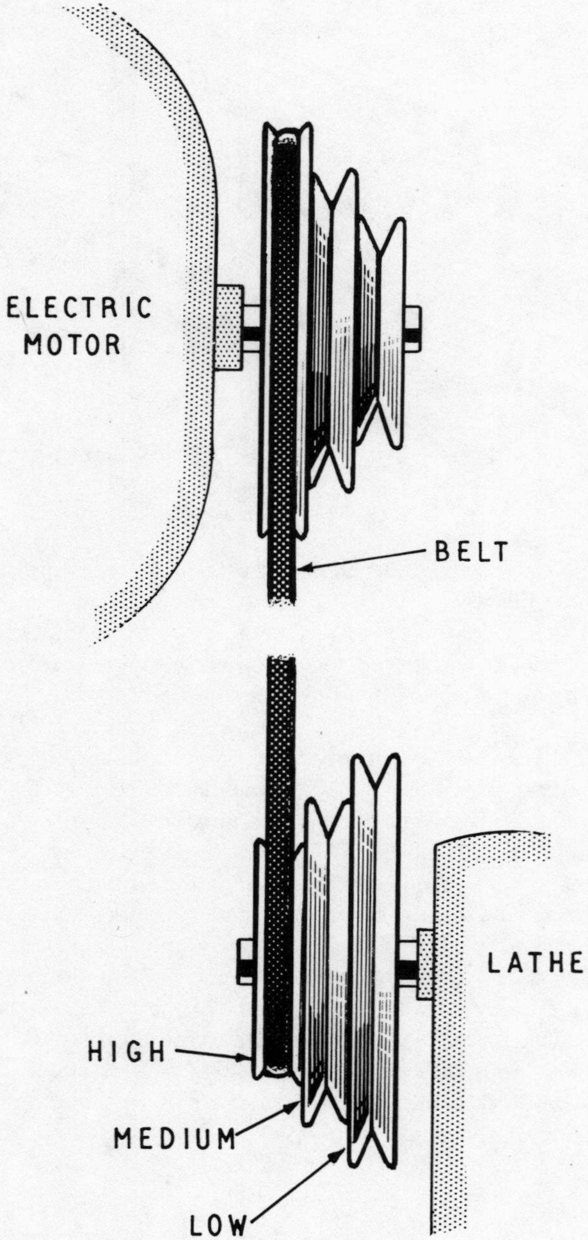

Fig. 69 Belting arrangement

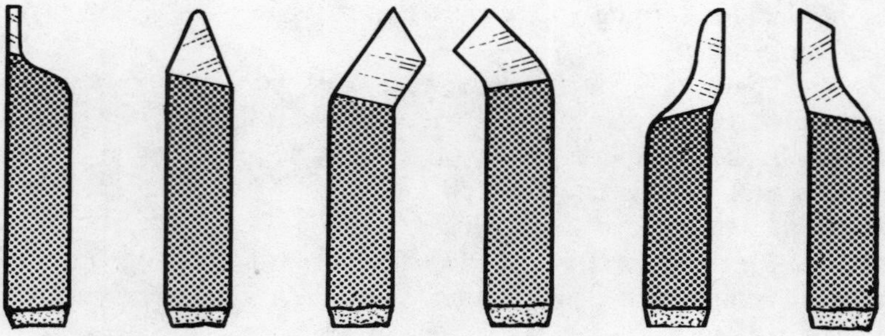

Fig. 70 Slide rest cutte

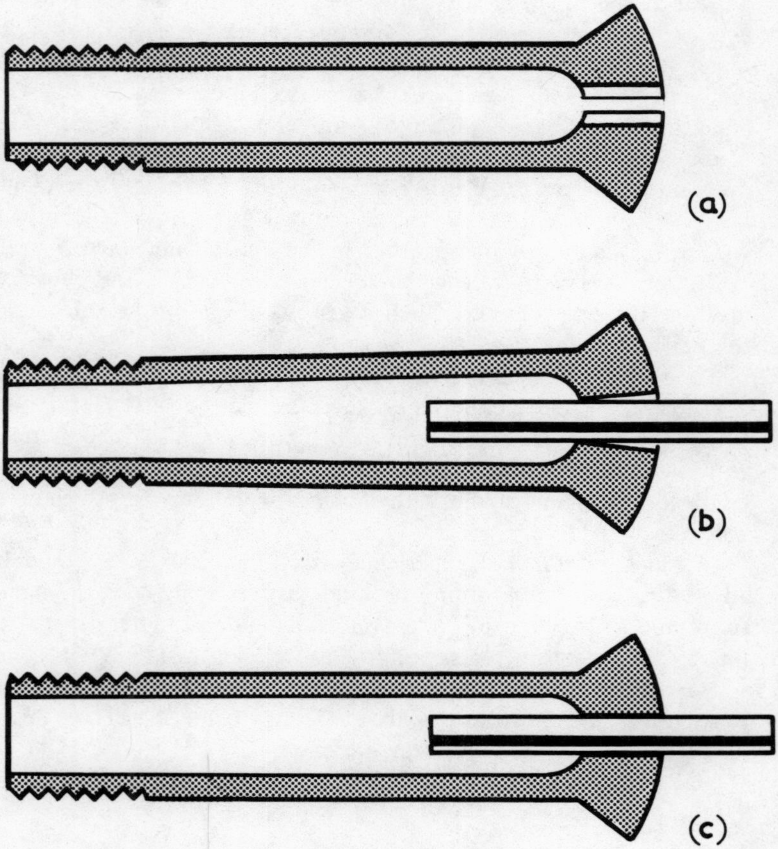

Fig. 71 Selecting a chuck

When turning steel a slow speed is required, but a high speed is needed for turning soft metals such as brass and for general polishing work.

When you have overhauled and set up your newly acquired secondhand lathe you will be anxious to put it in operation. Begin by gaining some experience on soft metal. Choose a length of brass rod about 2 in. long and between ⅛ in. and $\frac{3}{16}$ in. diameter, and select the correct size split chuck.

It will be seen from Fig. 71 (a) that the work is gripped in the conical head of the chuck over a short distance only. The ideal chuck size for any job is the one whose center hole diameter corresponds exactly with the diameter of the work, but this rarely happens. The range of sizes however is such that the slight differences that arise are of little consequence. But it is important to remember that a chuck should never be used if the diameter of the center hole is less than the diameter of the work. Fig. 71 (b) illustrates the effect of forcing open a chuck. Only the rear edge of the chuck jaws grip the work, and subsequent tightening of the draw-in spindle causes further straining and eventually the chuck fails to function correctly. Any attempt to turn work held in this way will cause the work to move in the chuck and not turn true.

Frequently it is necessary to reverse the work in the chuck to machine the opposite end. Such a task would be impossible under these conditions.

The remedy is never to use a chuck that is too small for the work, but to always select either the exact size or the next size larger. When the work is inserted into the chuck it should be parallel with the jaws as seen in Fig. 71 (c), and it will be found that tightening the draw-in spindle with finger tip pressure is sufficient to hold the work firmly.

If you have not had experience using the turns you would be advised to start your turning between centers fitted into the headstock and tailstock using a graver. It would be better to master this technique before attempting to use a slide rest.

Having selected the correct split chuck, push in the piece of brass rod so that it protrudes about ½ in. Locate the chuck in the headstock and tighten the draw-in spindle.

Make sure the belt is on the correct pair of pulleys to produce fast speed, and check that the lathe will be driven in a counter-clockwise direction when viewed from the tailstock. Switch on the machine and turn the diameter and the face as described for the turns.

When using the slide rest the angle of cut and height of the cutter are important, and in the beginning some trial and error is inevitable if one is to gain experience. Do not be tempted to take a big bite at the metal; it usually ruins the work and means having to begin again.

The big advantage of the slide rest is the guarantee that when one of the slide lead screws is turned, the cutter will move in a straight line without relying on the skill of the operator.

After turning a few pieces of brass, alter the belting to give a lower speed and try turning a piece of steel. The same cutting tools can be used but the process will take a little longer.

If you ever consider purchasing a new lathe two good manufacturers are:

F. W. DERBYSHIRE, INC.,
265 BEAR HILL ROAD,
WALTHAM, MASS. 02154

BALDOR ELECTRIC COMPANY,
P.O. BOX 6238
FORT SMITH, ARKANSAS 72901

both of whom would be pleased to send you literature.

V

Gearing

GEARING IS A MEANS OF CONVEYING MOTIVE POWER, AND IT IS used in watch and clock movements in the form of a gear train to transmit the energy of a mainspring or weights to the escapement.

Watch and clock manufacturers make every effort to keep frictional resistance to a minimum, and therefore give considerable thought to the shape of gear teeth.

The ideal arrangement would be to use discs placed edge to edge. This would certainly keep surface friction to a minimum, but slipping would undoubtedly take place so the drive would not be positive and regular. Hence we have to accept the necessity of using teeth.

The aim of a designer is to shape the teeth in such a way that when two gears are functioning together, the engaging teeth operate with a rolling action. In practice some sliding between the faces of teeth does take place but in well designed and correctly depthed gears it is kept to a minimum.

A wheel is usually the driver and a pinion is usually the driven. An imaginary straight line drawn through the pivots of a wheel and a pinion in engagement is known as the line of centers. If after a wheel tooth has made contact with a pinion leaf, some sliding of the mating faces takes place before the line of centers, the resistance created is called engaging friction. Any sliding after the line of resistance is known as disengaging friction.

The resistance of engaging friction is high and everything possible is done to keep it to a minimum, but disengaging friction creates a much lower resistance and is usually acceptable.

If we think again of two rotating discs placed edge to edge, one representing a wheel and the other a pinion, the two circumferences would represent the pitch circles of the two gears, Fig. 72.

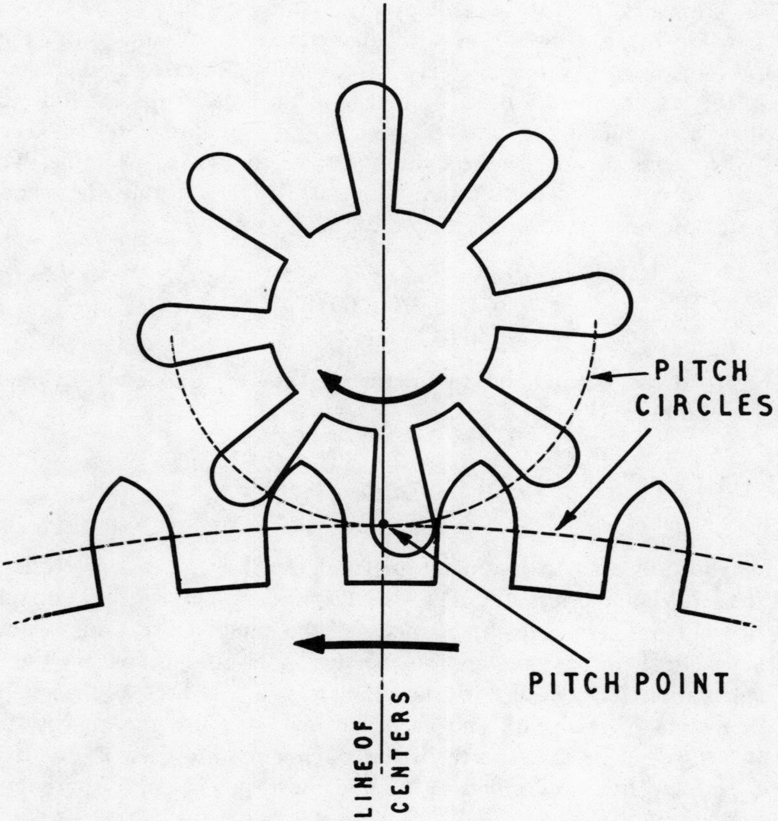

Fig. 72 Pitch circles

The ideal arrangement would be when contact between a wheel tooth and a pinion leaf is made at the pitch point. This occurs where the two pitch circles meet and are intersected by the line of centers. Until that point had been reached, the outgoing tooth would be driving with the minimum of disengaging friction.

When a wheel has 20 or more teeth it is known as a wheel, but if there are 19 or less teeth it is called a pinion and the teeth are referred to as leaves. In addition to the numbers of teeth and leaves, the ratio or proportion of wheel size to pinion size plays an important part in good depthing.

If we consider two discs with diameters of 1 in. and ⅛ in. the ratio of one to the other is expressed as 8:1. Because the diameter of the larger disc is 8 times greater than the smaller disc, it follows that the circumference must also be 8 times greater.

Problems arise when calculating for teeth. If, for example, we were to cut the teeth to a depth of 0.015 inches in the wheel and in the pinion, the new ratio would be:

$$\begin{aligned} 1\text{ in.} - (0.015\text{ in.} \times 2)&:\tfrac{1}{8}\text{ in.} - (0.015\text{ in.} \times 2) \\ &= 0.970:0.095 \\ &= 10.2\ :1 \end{aligned}$$

If on the other hand we increased the two diameters by this amount the ratio would be:

$$\begin{aligned} 1\text{ in.} + (0.015\text{ in.} \times 2)&:\tfrac{1}{8}\text{ in.} + (0.015\text{ in.} \times 2) \\ &= 1.030:0.155 \\ &= 6.6\ :1 \end{aligned}$$

The only way to retain the original ratio of 8:1 when altering the two diameters is to add to the larger disc 8 times the amount that is being added to the diameter of the smaller disc. This is not a practical proposition and so the designers compromise in their calculations. They consider the diameters of the discs as the pitch diameters of the gears and calculate the pitch circumferences. If the wheel is to have 80 teeth the pitch circumference is divided by 80 and then multiplied by 83 and the result is the full circumference of the wheel.

Similarly, pinion sizes are calculated in exactly the same way but only 1 leaf is added.

This addition of 3 teeth for a wheel and 1 leaf for a pinion is the compromise that designers make to overcome the difficulty in maintaining ratios.

These calculations can be more simply expressed as:

$$\text{full diameter} = \frac{\text{pitch diameter} \times \pi}{\text{number of teeth}} \times \frac{\text{number of teeth} + 3}{\pi}$$

where the Greek symbol π (pronounced pi) = 3.1416 which is the number of times the diameter of any circle can be divided into its circumference.

From this equation the pitch diameter of any wheel can be calculated if the full diameter and number of teeth are known:

$$\text{pitch diameter} = \frac{\text{full diameter} \times \pi}{\text{number of teeth} + 3} \times \frac{\text{number of teeth}}{\pi}$$

This equation also applies to pinions except that instead of +3 teeth the calculations include +1 leaf.

The combination of a wheel with 120 teeth and a pinion with 12 leaves produces very little engaging friction, but when the number of teeth and leaves are reduced there is an increase in resistance caused by an increase in engaging friction. This is very pronounced when pinions of 6 leaves engage a wheel with 60 teeth.

A well-matched wheel and pinion will cause trouble if the depth is incorrect. Correct depth exists when the two pitch circles just touch. If they overlap, Fig. 73 (a), the depth is too deep, and if the pitch circles do not touch the depth is too shallow, Fig. 73 (b).

With the majority of clocks it is not difficult to examine the depth of gearing because the wheels are usually accessible. Hold the pinion with the pointed end of a pegwood stick, and with another pegwood stick try the shake of the wheel. A light touch and a little experience will quickly produce the ability to assess whether or not the depth is correct.

It is somewhat different with a watch movement because the wheels are less accessible. When the movement has been disassembled, place between the plates the two wheels that are to be examined. Hold the pinion by pressing on the end of the pivot with a pointed pegwood stick, and with another pointed pegwood

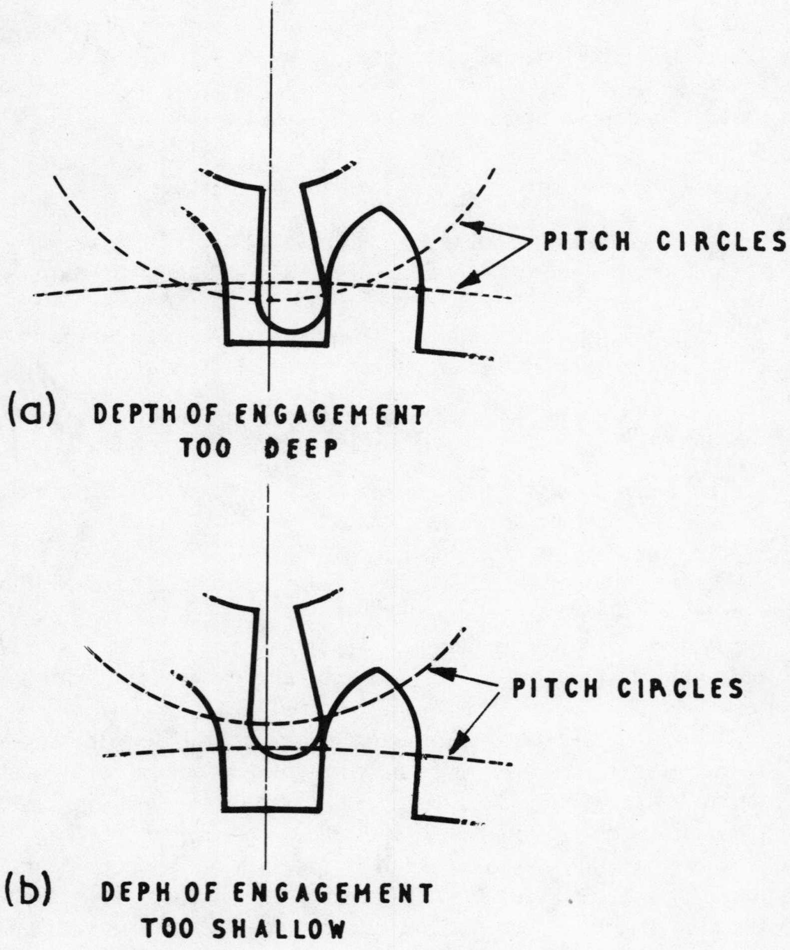

Fig. 73 Depth of engagement

stick lead the wheel round and study the depth. If this is not possible a depth tool, Fig. 74, will have to be used.

Let us assume it is the depth between the third wheel and the pinion of the fourth wheel that is in some doubt. Remove the two wheels from the movement and set them up in the depth tool, Fig.

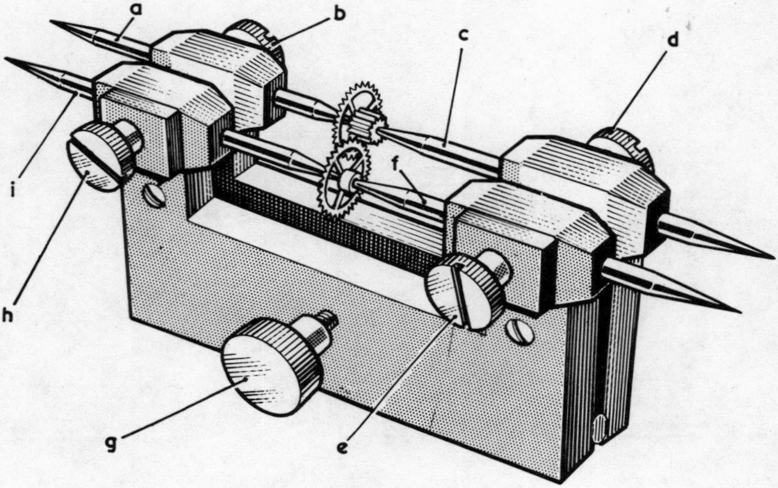

Fig. 74 Depth tool

75. The runners should be positioned so that the female ends are pointing inward to receive the wheel pivots, leaving the pointed ends of the runners to face outward.

The setting up procedure is as follows:

1. tighten thumbscrew (b)
2. place the pointed end of runner (a) in the third wheel pivot hole and keep the tool held perfectly upright.
3. adjust screw (g) until the pointed end of runner (i) is brought into line with the pivot hole of the fourth wheel. Adjust runner (i) so that the pointed end enters the pivot hole.
4. tighten thumbscrew (h)
5. with the points of runners (a) and (i) in the pivot holes of the third and fourth wheels, check that the tool is square with the movement and not leaning over. Adjust if necessary.
6. position the two wheels between the female ends of the runners, and adjust runners (c) and (f) so that the third wheel engages at the end of the fourth wheel pinion.

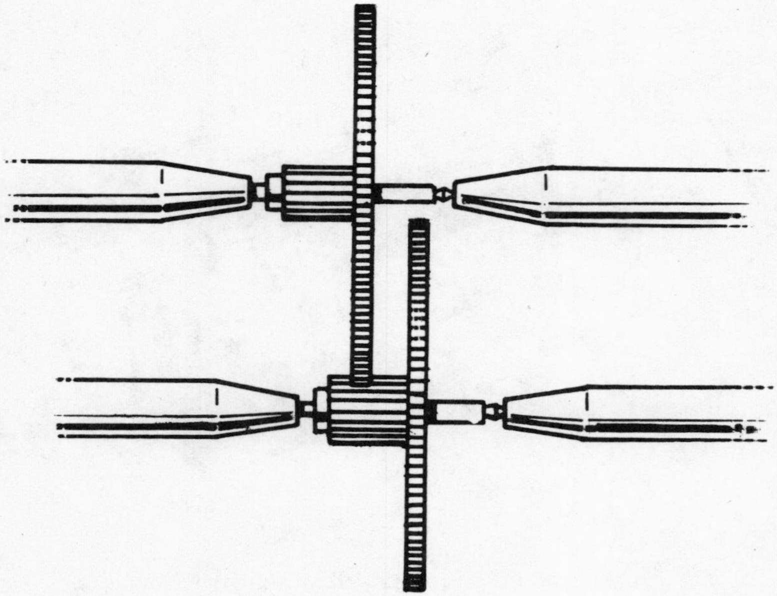

Fig. 75 Checking depth of mesh

Tighten thumbscrews (d) and (e) and put a little extra pressure on the end of the fourth wheel pivot so that it offers some resistance to being turned.

7. rotate the third wheel by finger and study the depth of engagement with the fourth wheel pinion through a glass. Reference to Fig. 73 should help in forming an opinion.

The remedies for incorrect depthing in clocks and watches are:

DEPTH TOO DEEP	DEPTH TOO SHALLOW
1. reduce wheel diameter in topping tool	1. stretch the wheel
or	or
2. fit smaller wheel or pinion	2. fit larger wheel or pinion

First, however, we should check the ratio of the wheel to the pinion by using a sector, Fig. 76.

A sector consists of two identical, but handed, graduated rules hinged at the bottom and free to open within the limit of the slotted plate at the top.

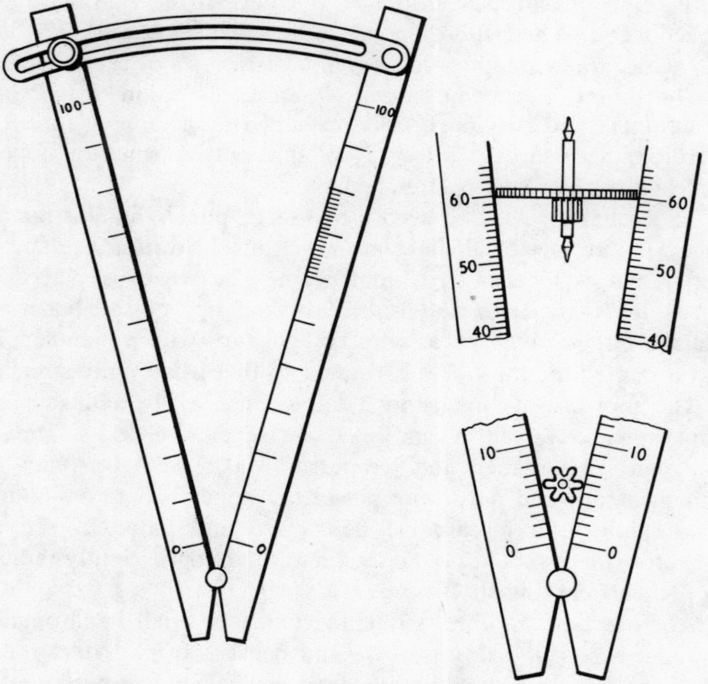

Fig. 76 A sector

As an example we will consider a wheel with 60 teeth that engages with a pinion of 6 leaves. The arms of the sector are opened until the wheel is supported between the two arms opposite the 60 mark. Make sure the wheel is being held at its maximum diameter, i.e., at the tops of the teeth. The locking screw is now tightened to prevent the arms from moving.

The pinion of the other wheel is now positioned in the sector with its maximum diameter against the graduated edges. If the pinion rests against the 6 mark on each arm, the size of the pinion is correct, but if it rests above or below the 6 mark then a replacement pinion is the best remedy.

If a pinion is too big in diameter the leaves will be too far apart at the ends and in most instances will have worn a groove in the wheel teeth. When this happens a new wheel should be fitted as well as a new pinion.

The action of stretching a wheel causes distortion to the shape of the teeth and they have to be reshaped on a topping tool. It is therefore necessary to select a suitable cutter from the topping tool before the wheel is stretched.

To stretch a wheel we need a two-piece punch, as illustrated in Fig. 77. The lower half has flats which are held in a vise and the upper half is located by a center pin. The wheel is placed between the two halves and is held by the tips of the finger and thumb. The top is given a series of light taps with a hammer, and at the same time the wheel is rotated so that it is evenly stretched.

The tool can be made on a lathe quite easily and suggested dimensions are given in Fig. 78. Use a silver steel rod so that the tool can be hardened and tempered. Make the lower half first and then turn and polish the pin of the upper half to a sliding fit in the hole. All faces must be square and parallel. The two operating faces should be polished and their edges slightly radiused to prevent marking the wheels.

Another type of wheel stretching tool is a small hand-operated machine which is held in a vise and consists of two steel rollers. The wheel to be stretched is placed between the rollers, and an adjusting screw brings the rollers together gripping the band of the wheel. The rollers are turned by a small handle and at the same time the operator guides the wheel through the rollers. Because the stretching takes place on the band of the wheel the teeth retain their original shape and there is no need for topping.

A topping tool cutter is shown in Fig. 79 (a). The correct method of selecting a cutter that will produce the same profile and size of the wheel teeth is to insert the cutter between two teeth and then withdraw it sideways in a cutting attitude. If the cutter

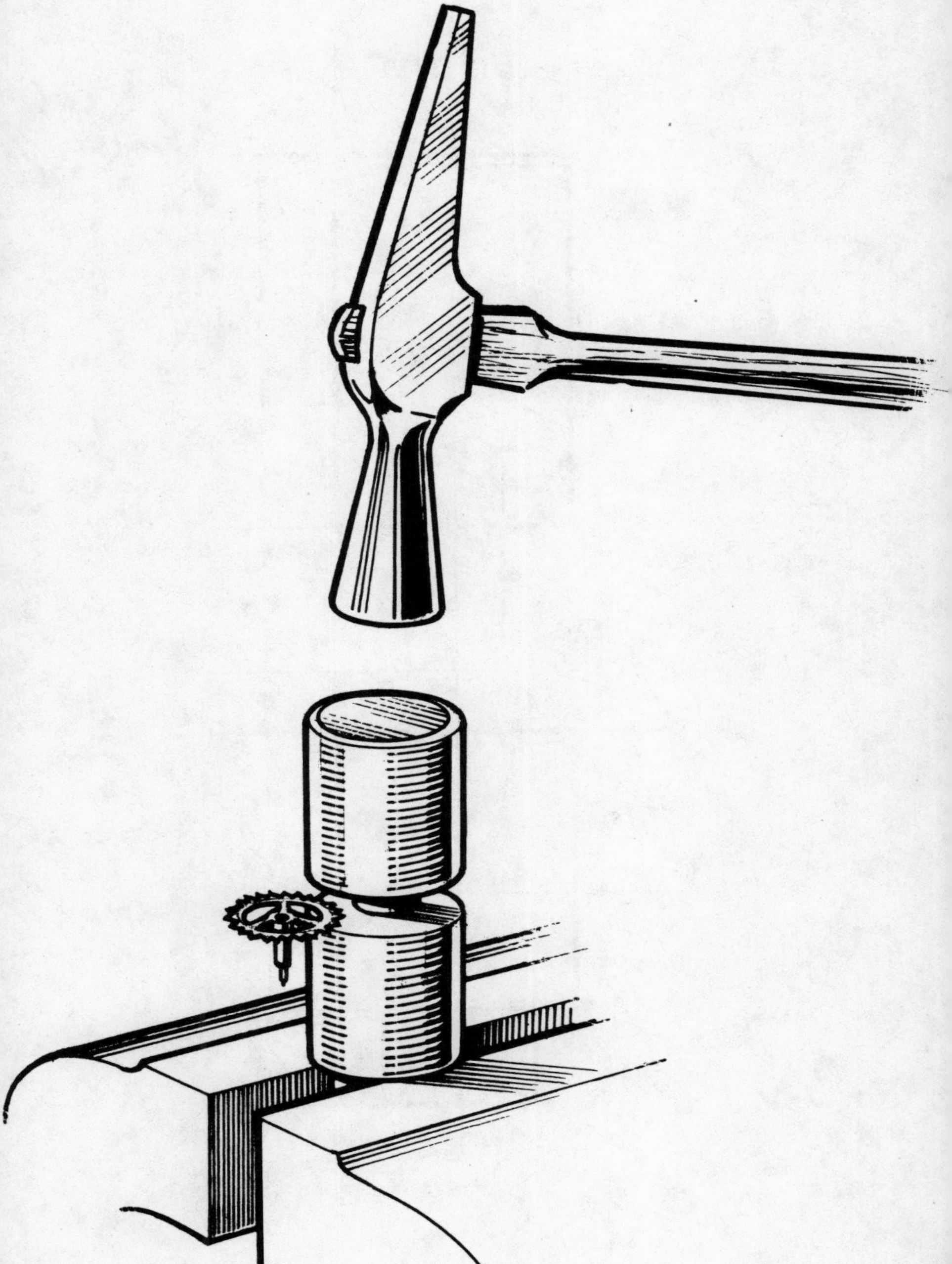

Fig. 77 Wheel stretching

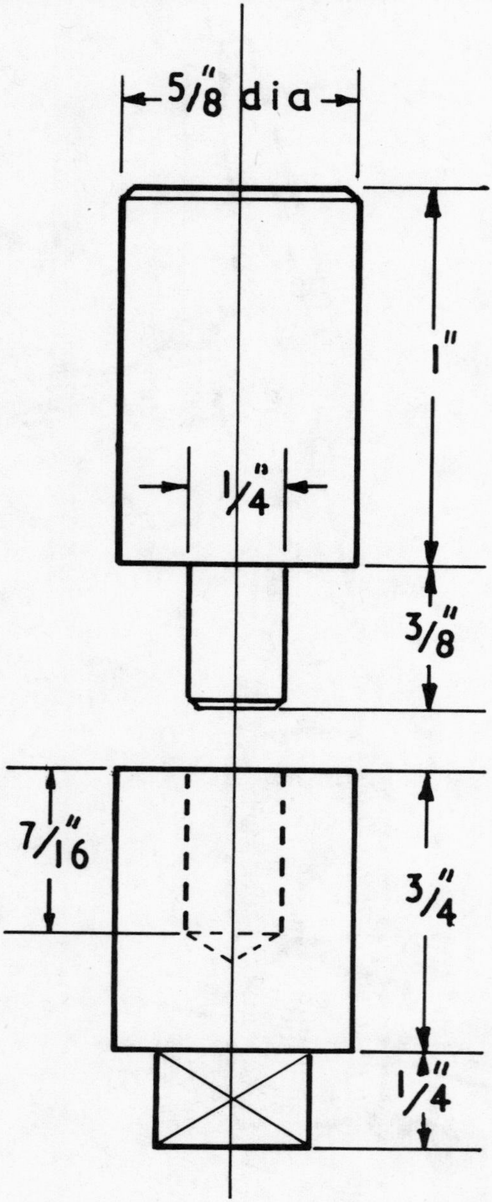

Fig. 78 Wheel stretching punch

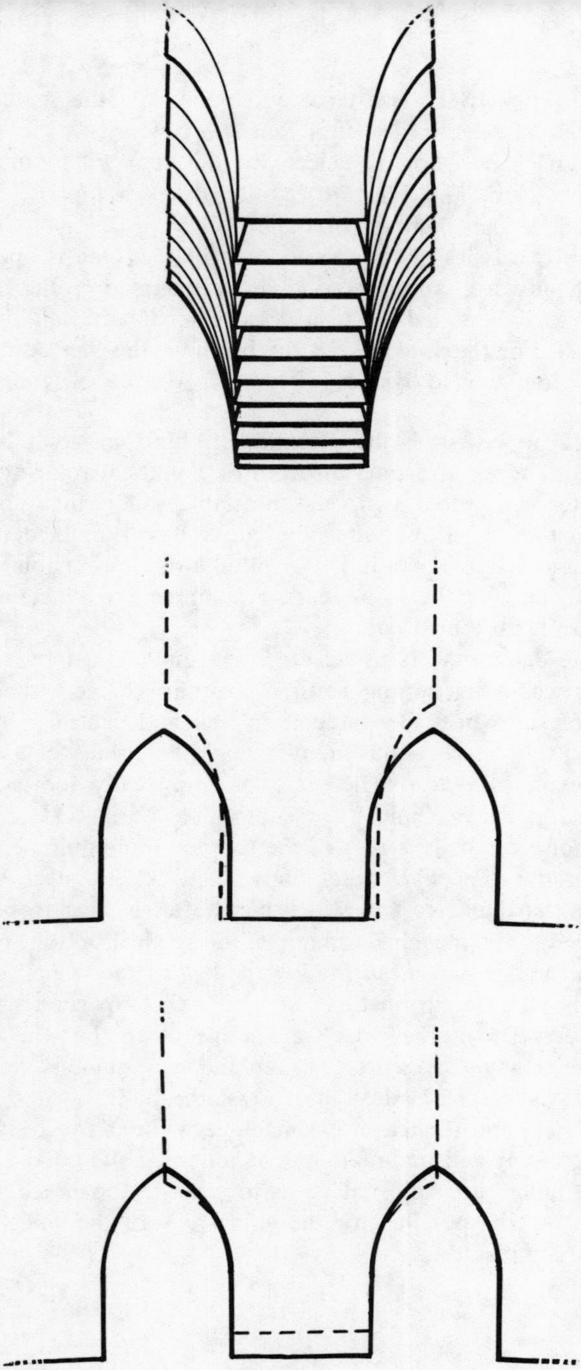

Fig. 79 Wheel topping cutter

is pushed down between two teeth and pulled out the same way, there is a risk of selecting a cutter that is too wide.

Fig. 79 (b) shows how to select a cutter that will reduce the width of the teeth, and Fig. 79 (c) illustrates a topping cutter that will reduce the diameter of the wheel.

There are two types of topping tools available. One is operated by a large hand wheel and gut string and is mounted on the bench, and the other is supplied as an accessory to a lathe and fits on the slide rest. For the student and the beginner the hand operated model is recommended because a greater degree of control is possible.

The wheel to be cut is fitted to the machine horizontally and is rotated, one tooth at a time, in front of a vertical rotary cutter.

The cutters are fitted with an adjustable guide that engages the teeth of the wheel and when the cutter has completed its cut the guide comes into operation. Its spiral formation enables it to engage the next gap between teeth and move the wheel round in line ready for the next cut.

When the cutter has been selected, the guide must be so adjusted that when the cutting teeth are sitting correctly between two teeth of the wheel the guide is in line and central with the next gap on the wheel. Adjustment is made by means of a screw on the side of the cutter. The cutter is now ready for use and can be fitted to the machine.

A range of brass beds is part of the topping tool equipment and they are used rather like watch movement rests. Select a bed with the largest diameter that will leave the teeth clear to be cut.

Fit the bed to the machine and adjust the vertical bottom runner up through the bed and on to the lower pivot of the wheel so that the wheel is just resting on the bed. Bring the top runner down to the upper pivot of the wheel and adjust it so that the wheel arbor is under slight pressure between the two runners, and the face of the wheel rubs against the end of the bed.

If the wheel is positioned in the machine without any resistance to turn, the cutter will undoubtedly cut too wide. If, on the other hand, the runners are adjusted so as to provide too much resistance, it will not be possible for the guide to turn the wheel after each cut is completed.

An adjusting screw is provided to bring the wheel to the correct height for cutting. An indicator on the machine marks the position for the center of the thickness of the wheel.

Another screw adjusts the position of the cutter in relation to the wheel. The cutter must move along a line that passes through the center of the wheel or the teeth will be cut at an angle and will be leaning over.

A third adjusting screw limits the distance to which the cutters can travel, and so controls the depth of the cut.

By operating an arm on the side of the machine, the cutter is moved up to the wheel. Examine the entry of the cutter into the wheel very closely and if necessary move the wheel round so that the cutter enters centrally between the teeth.

Operate the machine slowly, keeping the cutter up to the wheel and observe the action of the guide when the cut is complete. It may be necessary to alter slightly the adjustment of the guide.

Continue to operate the machine slowly, and always make sure that the guide has moved the wheel round before making another cut.

The count of a gear train is the number of single vibrations per hour made by the pendulum of a clock, (usually referred to as the number of vibrations per minute), or the number of single beats per hour made by a watch or clock balance.

It is frequently helpful to know the count of a train, for example, when selecting and fitting a replacement balance spring, renewing a pendulum, using a timing machine, or calculating the size of a lost wheel.

To calculate the count we multiply the number of teeth in the center, third and fourth wheels, and twice the number of teeth in the escape wheel, and divide by the number of leaves in the third, fourth and escape wheel pinions. We must multiply the number of teeth in the escape wheel by two, because each tooth acts on the two pallets separately, moves the lever twice, and causes two beats of the balance or pendulum.

Example 1. Consider a watch movement with the following train:

center wheel	64
third pinion	8

third wheel	60
fourth pinion	8
fourth wheel	70
escape pinion	7
escape wheel	15

By applying the formula we have:

$$\frac{64 \times 60 \times 70 \times (15 \times 2)}{8 \times 8 \times 7} = 18{,}000$$

This means that the balance will beat 18,000 times in one hour or 300 times per minute. Each beat is the swing of the balance or pendulum from one side to the other.

If a wheel complete with its pinion is lost from a watch or clock movement the normal procedure is to fit a manufacturer's replacement. This should not prove difficult with a watch unless it is a very old timepiece. With clocks it can be troublesome because there are many movements in existence for which replacement parts are unobtainable. Under these circumstances it will be necessary to have a new wheel cut, and we will need to know how to calculate the number of teeth in the wheel, the number of leaves in the pinion, and their respective diameters.

Example 2. Let us assume the fourth wheel and pinion has been lost from the watch movement that we used in Example 1.

Our equation was:

$$\frac{64 \times 60 \times 70 \times (15 \times 2)}{8 \times 8 \times 7} = 18{,}000$$

We know that the fourth wheel has 70 teeth and that its pinion has 8 leaves but when a wheel and pinion is lost we rarely know its size beforehand. So let us make some calculations and see if we arrive at these figures of 70 and 8.

If we substitute the letter x for the fourth wheel and the letter y for the fourth pinion and rewrite the equation we have:

$$\frac{64 \times 60 \times x \times (15 \times 2)}{8 \times y \times 7} = 18{,}000$$

Before, we proceed, however, one might ask how do we know that the count is 18,000 if we are not in possession of all the wheels and pinions from which our calculation can be made.

There are other methods of obtaining the count apart from using the number of teeth and leaves in the train.

The beats of a balance can be counted physically by suspending the balance above a smooth hard surface, setting the balance in motion and, over a known period of time, counting the number of taps the staff will make when the hairspring unwinds at every alternate beat.

The number of vibrations made by a clock pendulum can be obtained from calculations based on the measured length of the pendulum and reference to Chapter 18 will give you the detailed procedure.

Returning to the equation we had:

$$\frac{64 \times 60 \times x \times (15 \times 2)}{8 \times y \times 7} = 18{,}000$$

which when simplified becomes:

$$\frac{14400x}{7y} = 18{,}000$$

$$14400x = 18{,}000 \times 7y$$

$$x = \frac{18{,}000 \times 7y}{14{,}400}$$

$$x = 8.75y$$

Our calculations have shown that wheel x = 8.75 times pinion y. This means:

(a) the gear ratio is 8.75:1:
(b) wheel x has 8.75 times as many teeth as pinion y has leaves
(c) wheel x has a pitch diameter 8.75 times greater than that of pinion y
(d) wheel x rotates once to every 8.75 revolutions of pinion y.

This we know to be correct because the missing wheel had 70 teeth and the pinion had 8 leaves.

The center wheel of a watch movement carries the minute hand and therefore rotates once every hour. The fourth wheel rotates 60 times every hour (1 revolution per minute) and is used to carry a small second hand when fitted.

In our example the center wheel has 64 teeth and the third pinion has 8 leaves, a ratio of 8 : 1. Because the center wheel rotates once every hour, the third pinion and wheel must rotate 8 times as fast, i.e., 8 times every hour.

If the third wheel, which rotates 8 times each hour, is to drive the fourth pinion and wheel 60 revolutions every hour the ratio will have to be 7.5 : 1 ($8 \times 7.5 = 60$).

We know the third wheel has 60 teeth, and to bring about a ratio of 7.5 : 1 the fourth pinion will have to have 8 leaves.

Previous calculations have shown that the fourth wheel must have 8.75 times as many teeth as the pinion has leaves, and so the wheel must have 70 teeth.

Example 3. With the same watch movement assume that the third wheel and pinion is lost. Our equation will now be:

$$\frac{64 \times x \times 70 \times (15 \times 2)}{y \times 8 \times 7} = 18{,}000$$

and by simplification $x = 7.5y$.

The center wheel of 64 teeth rotates once every hour and the fourth pinion of 8 leaves rotates 60 times every hour. Between the two wheels there is a ratio of 60:1 in two stages. We can account for a ratio of 8:1 from the 64 teeth of the center wheel and the 8 leaves of the third pinion. To make up the difference we need a ratio of 7.5 : 1 ($8 \times 7.5 = 60$).

The third wheel and pinion must therefore have a ratio of 7.5 : 1 and so the two stages will be:

center wheel 64, third pinion 8 = 8:1
third wheel 60, fourth pinion 8 = 7.5:1
resultant ratio 60:1

Similar calculations can be made with clock movement trains, except that clock movements do not have fourth wheels and pinions.

In addition to calculating the number of teeth and leaves in the missing third wheel and pinion, we can also calculate the diameter of the wheel and of the pinion. This information will certainly be required if a new wheel is to be cut.

Measure the diameter of the center wheel, making quite sure that the measurement is taken at the tips of the teeth to give the correct reading. The best instrument for this is probably the micrometer.

Now measure the distance between the center wheel jewel hole and the third wheel jewel hole. This can be done by adjusting the male runners of a depth tool to the holes, and then transferring the setting to a slide gauge.

Example 4. Fig. 80. We will assume that the full diameter of the center wheel is 0.352 in. and the distance between the two holes is 0.190 in. With this information we can begin our calculations.

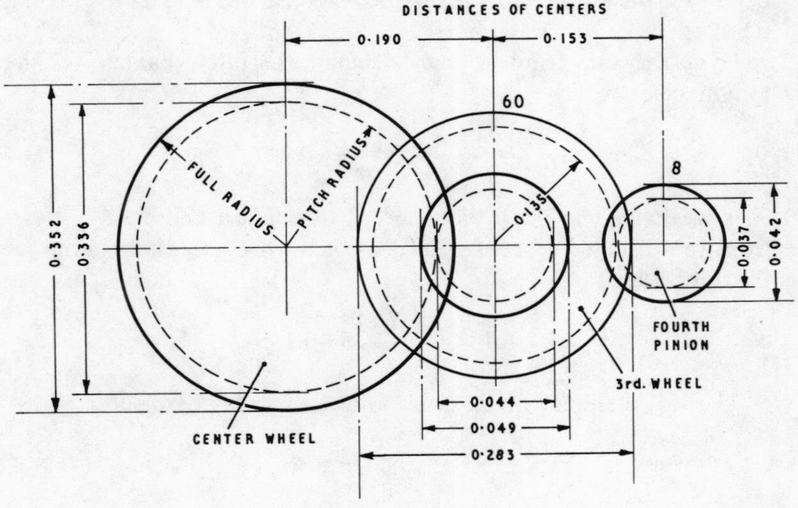

Fig. 80 Distances of centers

The pitch diameter of the center wheel is:

$$\frac{0.352 \times \pi}{67} \times \frac{64}{\pi} = 0.336 \text{ in.}$$

The pitch radius of the center wheel + the pitch radius of the third pinion = the distance measured between pivot holes = 0.190 in. Therefore the pitch diameter of the center wheel + the pitch diameter of the third pinion = 0.190 in. × 2 = 0.380 in.

Hence, the pitch diameter of the third pinion = 0.380 − 0.336 = 0.044 in. and so the full diameter of the pinion must be:

$$\frac{0.044 \times \pi}{8} \times \frac{9}{\pi} = 0.049 \text{ in.}$$

To find the full diameter of the third wheel we need to measure the distance of centers between the third and fourth wheel holes. Let us assume that the distance is 0.153 in., which we know is the sum of the third wheel and the fourth pinion pitch radii. By multiplying this figure by 2 we have the two pitch diameters.

We know that the third wheel has 60 teeth and the fourth pinion has 8 leaves, and we now have to measure the full diameter of the fourth pinion, which for this exercise we will take as being 0.042 in.

From this dimension we can calculate the pitch diameter of the pinion which will be:

$$\frac{0.042 \times \pi}{9} \times \frac{8}{\pi} = 0.037$$

If we halve the pitch diameter of the pinion and take it away from the distance of centers, we are left with the pitch radius of the third wheel:

$$0.153 - \frac{0.037}{2} = 0.135$$

The full diameter of the third wheel can now be calculated as:

$$\frac{0.135 \times 2 \times \pi}{60} \times \frac{63}{\pi} = 0.283 \text{ in.}$$

The gears behind a watch or clock dial are called the motion work. Their function is to provide a drive for the hour and minute hands at a reduced ratio, Fig. 81.

The center wheel rotates once in every hour. Mounted on one end of the center arbor is the cannon pinion which carries the

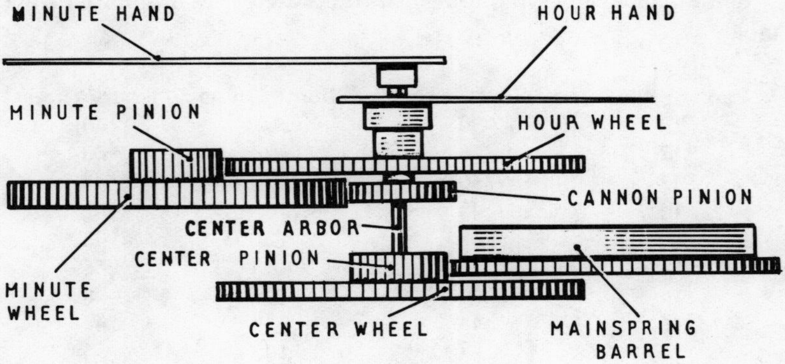

Fig. 81 Motion work

minute hand. The cannon pinion also drives the minute wheel and the minute pinion drives the hour wheel. The hour wheel, which carries the hour hand, rotates once in every 12 hours and so between the cannon pinion and the hour wheel there must be a reduction of 12:1. This is arranged in two stages:

Cannon pinion	10
Minute wheel	40
Minute pinion	12
Hour wheel	36

or

Cannon pinion	12
Minute wheel	36
Minute pinion	10
Hour wheel	40

$$= \frac{40 \times 36}{12 \times 10} = 12$$

Some typical clock motion work trains are given in the following table:

Cannon Pinion	Minute Wheel	Minute Pinion	Hour Wheel
10	30	10	40
12	36	10	40
12	36	12	48
14	40	10	42
14	42	12	48
14	42	14	56
16	40	10	48
16	48	12	48
16	48	14	56
16	48	16	64
18	48	12	54
18	54	14	56
18	54	16	64
20	60	18	72

The subject of clock pendulums is covered in Chapter XVIII, but the following table is included in this chapter because of the reference to gear trains:

CENTER WHEEL	THIRD WHEEL PINION	THIRD WHEEL	ESCAPE WHEEL PINION	ESCAPE WHEEL	PENDULUM VIBRATIONS PER MINUTE	PENDULUM LENGTH IN INCHES
128	16	120	16	30	60	39.14
112	14	105	14	30	60	39.14
96	12	90	12	30	60	39.14
80	10	75	10	30	60	39.14
64	8	60	8	30	60	39.14
68	8	64	8	30	68	30.49
70	8	64	8	30	70	28.75
72	8	64	8	30	72	27.17
75	8	60	8	32	75	25.05
72	8	65	8	32	78	23.15
75	8	64	8	32	80	22.01
84	8	64	8	30	84	19.97
86	8	64	8	30	86	19.06
88	8	64	8	30	88	18.19
84	7	78	7	20	89.1	17.72
80	8	72	8	30	90	17.39
84	7	78	7	21	93.6	16.08
94	8	64	8	30	94	15.94
84	8	78	8	28	95.5	15.45
108	12	100	10	32	96	15.28
84	9	84	8	30	98	14.66
84	7	78	7	22	98	14.66
84	8	78	8	29	98.9	14.41
80	8	80	8	30	100	14.09
85	8	72	8	32	102	13.54
84	8	78	8	30	102.4	13.44
84	7	78	7	23	102.5	13.40
105	10	100	10	30	105	12.78
84	8	78	8	31	105.8	12.59
84	7	78	7	24	107	12.30
96	8	72	8	30	108	12.08
84	8	78	8	32	109.2	11.82
88	8	80	8	30	110	11.64
84	7	77	7	25	110	11.64
84	7	78	7	25	111.4	11.35
84	8	80	8	32	112	11.22
84	8	78	8	33	112.2	11.15
96	8	76	8	30	114	10.82
115	10	100	10	30	115	10.65

CENTER WHEEL	THIRD WHEEL PINION	THIRD WHEEL	ESCAPE WHEEL PINION	ESCAPE WHEEL	PENDULUM VIBRATIONS PER MINUTE	PENDULUM LENGTH IN INCHES
84	7	78	7	26	115.9	10.49
96	8	80	8	30	120	9.78
84	7	70	7	30	120	9.78
84	7	78	7	27	120.3	9.73
90	8	84	8	31	122	9.46
84	7	78	7	28	124.8	9.02
100	8	80	8	30	125	9.01
90	8	84	8	32	126	8.87
100	10	96	10	40	128	8.59
84	7	78	7	29	129.3	8.42
100	8	78	8	32	130	8.34
84	7	77	7	30	132	8.08
84	7	78	7	30	133.7	7.90
90	8	90	8	32	135	7.73
84	7	78	7	31	138.2	7.38
84	8	80	8	40	140	7.18
120	8	71	8	32	142	6.99
84	7	78	7	32	142.6	6.93
100	8	87	8	32	145	6.69
84	7	78	7	33	147.1	6.50
100	8	96	8	30	150	6.26
84	7	78	7	34	151.6	6.10
96	8	95	8	32	152	6.09
84	7	77	7	35	154	5.94
104	8	96	8	30	156	5.78
84	7	78	7	35	156	5.78
120	9	96	8	30	160	5.50
84	7	78	7	36	160.5	5.47
84	7	78	7	37	164.9	5.18
132	9	100	8	27	165	5.17
84	7	78	7	38	169.4	4.88
128	8	102	8	25	170	4.87
84	7	78	7	39	173.8	4.65
84	7	77	7	40	175	4.55
84	7	78	7	40	176	4.43

TABLE OF CLOCK TRAINS

VI

Cleaning Machines

THE ALTERNATIVE METHOD TO HAND BRUSH CLEANING IS BY MAchine. This method provides cleaning by fluid friction and produces excellent results in far less time than it takes to clean by hand. Some machines are designed to be operated manually while others are hydraulically operated and are automatic in action. Both are fitted with an electric motor at the head of the column for rotating the cleaning basket.

Fig. 82 illustrates an L & R Vari-Matic which is automatic in action. It is fitted with a turntable on which are mounted three glass jars and a drying chamber. Jar No. 1 is filled with cleaning solution and jars No. 2 & 3 contain rinsing fluid.

As with brush cleaning the movement is completely disassembled. The parts are placed in a wire gauze basket, the machine is switched ON and the start button is depressed. The cycle of operations is as follows:

1. the basket motor rises up the column and the turn-table brings jar No. 1 into position directly beneath the basket.
2. the basket is lowered into the cleaning solution where it rotates forward and backward alternately for approximately three minutes.
3. the basket rises to spin-off position where it spins in one direction for about seven seconds.

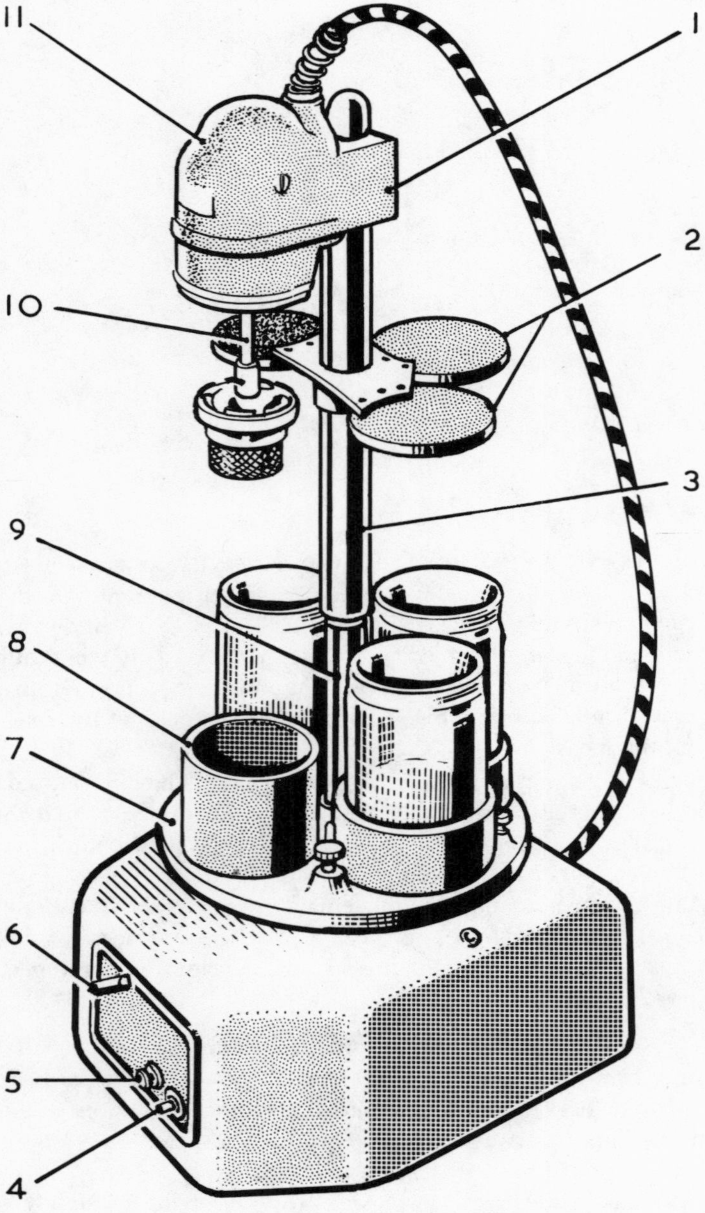

Fig. 82 Cleaning Machine

1	Locking screw	7	Turntable
2	Jar covers	8	Drying chamber
3	Outer column	9	Inner column
4	High-low switch	10	Motor shaft
5	Starting switch	11	Electric motor
6	Speed control		

4. the process is repeated for jars No. 2 & 3, and then the basket is lowered into the drying chamber where it rotates at a reduced speed in one direction only for about three minutes. The basket then rises and the motor cuts off.

The cleaning solution does not normally cause rust, but if it is allowed to dry on the parts it can cause staining which is not easy to remove.

The chemical composition of rinsing fluid can be affected by exposure to sunlight and is often the cause of rust. Many bench operators paint the outside of jars No. 2 and 3 with matte black paint to keep out the light.

Perhaps the best safeguard against rust is to remove all steel parts from the movement and clean them by hand in benzine. Certainly the balance spring must be dealt with in this way. Under no circumstances should the mainspring be put into the cleaning solution while coiled in the barrel. The spring must be removed from the barrel so that it can be thoroughly rinsed and dried.

Cleaning solution can soften the shellac which holds impulse pins and pallet stones in place. If the stones become loose they can easily be lost. It is much safer to wash the balance and the escapement in benzine.

After machine cleaning, all pivot and jewel holes should be pegged in the usual way, and normal oiling procedure is carried out.

In the upper half of the cleaning basket is a detachable partitioned tray into which the wheels and small pieces are placed. The plates and other large pieces are spread out on the bottom of the basket with the minimum of overlapping.

It is important to maintain the correct level of fluid in the jars. Excess liquid will overflow when the basket is rotating at normal speed, and too little fluid may not cover the parts to be cleaned and staining will result.

This method of mechanical cleaning is efficient and very popular. Such machines are in use in the majority of workshops and will continue to be used for many years to come.

Nevertheless, in terms of progress, this type of machine has become outdated by the introduction of the ultrasonic method.

Ultrasonic cleaning produces a standard of cleanliness higher than can be achieved by any other method. The equipment includes a transducer which is immersed in the cleaning fluid, and a valve oscillator. The oscillator generates electrical impulses which are passed to the transducer by flexible cable, and are then transformed into mechanical vibrations that set up high pitched sound waves. The pitch of the sound waves is too high to be audible by human ear and it is therefore known as ultrasonic.

These high frequency vibrations or sound waves taking place in the cleaning fluid produce countless microscopic bubbles which are bubbles of vapor of the cleaning fluid; a process known as cavitation.

These cavity bubbles adhere to the surface of anything which is immersed in the fluid until the built-up pressure causes the bubbles to implode. The continual rapid formation and collapse of these bubbles sets up an erosive action which dislodges and removes any dirt that is present, even in small orifices and blind holes.

The continued movement of sound waves passing across the surface of the fluid sweeps the dirt to the side of the container and holds it there allowing withdrawal of the cleaned components free from contamination.

Apart from speed and thoroughness, ultrasonic equipment enables delicate parts to be cleaned without damage. Another advantage is the ability to clean a movement without disassembling.

Fig. 83 illustrates an L & R ultrasonic unit which, when coupled with a cleaning machine, will enable the machine to clean by ultrasonic method. In this instance the generator illustrated has been made to suit the shape of the Vari-Matic because it is used as a base on which to stand the machine.

Ultrasonic units can be used in conjunction with any conventional cleaning machine, automatic or manual, and the operation of the cleaning machine remains unchanged.

The extraordinary effects of ultrasonic cleaning can be achieved only by using a chemical that has been specially formulated to take advantage of ultrasonic cavitation. By using L & R Duo-Lube watch lubricant diluted in L & R ultrasonic watch rinsing solution, a complete watch movement can be cleaned and lubricated in one

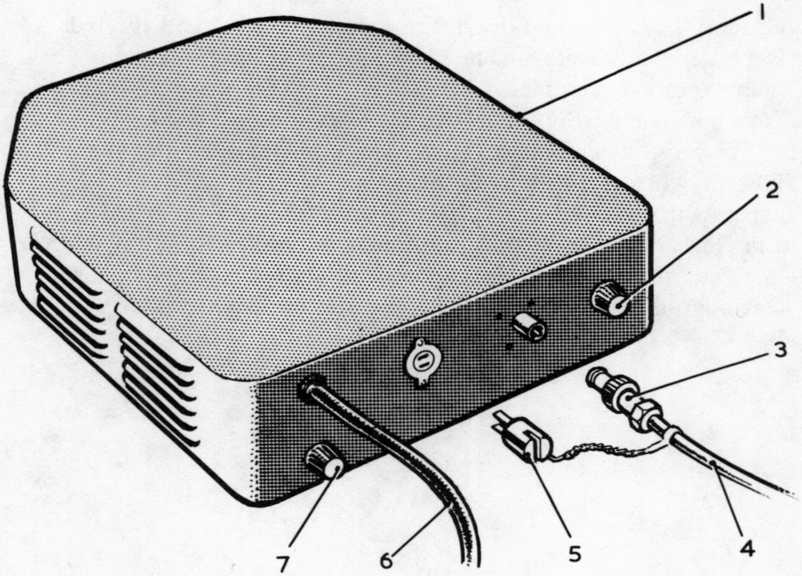

Fig. 83 Ultrasonic unit

1 Generator
2 Tuning knob
3 Coaxial connector
4 Cable to tank (if used)
5 By-pass plug
6 Transducer cable
7 Fuse holder

operation without disassembling. Only the dial and hands need be removed.

After cleaning and drying it will be necessary to lift the balance and the cock from the movement and dip them in hairspring cleaner. They are then dried, the jewels are oiled and the balance is refitted to the movement. The job of cleaning and oiling will then have been completed in far less time than it would otherwise have taken, and without the attendant possibility of accidents.

For large assemblies, such as clock movements, or where the flow of work is continuous, ultrasonic cleaning tanks which dispense with the need of the conventional basket type cleaning machine are available. A further refinement is to introduce a selector box into the system which allows the operator to select tank or cleaning machine, depending on the quantity and bulk of work in hand.

VII

Timing Machines

BASICALLY THERE ARE TWO TYPES OF ELECTRONIC TIMING MAchines. These are the visual and the recording, both of which can be used for watches and clocks fitted with mechanical or electrical movements.

In each case, the machines are supplied with a fully maneuverable stand carrying a built-in microphone to which a watch movement can be clipped. It is thereby possible to test the movement in any attitude.

When testing a clock movement the microphone is clipped as close as possible to the escapement.

The machine that produces the visual indication uses a cathode-ray tube. The tick of the movement being tested is picked up by the microphone and the electrical output is amplified through a loudspeaker in the timing machine. Monitoring headphones can be used if the loudspeaker proves to be a nuisance to others. At the same time, the tick is shown on the screen as a bright colored spot. If the movement is out of adjustment, the spot moves round the screen clockwise or counter-clockwise, depending on whether the error is a gain or loss. A knob is turned until the moving spot is brought to a standstill, and the amount of error can then be read from a dial.

In addition to this test, the behavior of the escapement can be displayed on the screen in the form of an oscillogram from which it is possible to diagnose escapement faults.

The recording machine, Fig. 84, prints each beat of the escapement on a moving roll of paper. The first beat on the microphone starts the recording device by means of the microphone amplifier, and when the movement is removed from the microphone the recording device stops automatically.

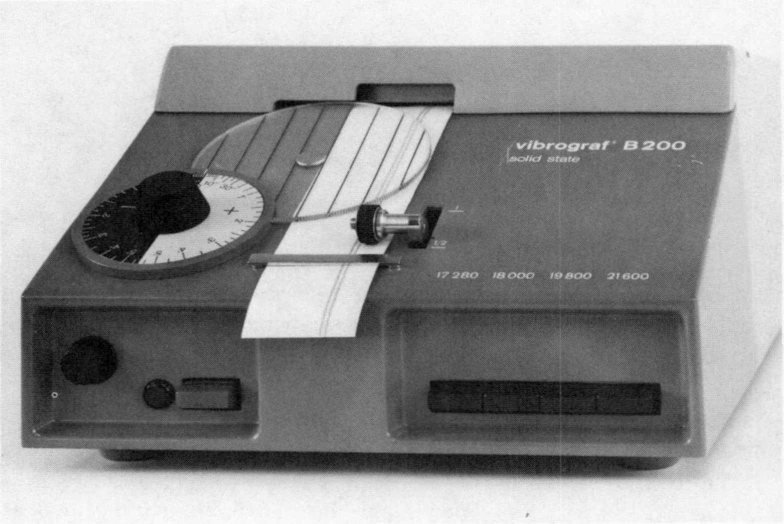

Fig. 84 Timing machine

An engraved transparent disc mounted above the recording tape can be rotated. When the machine is recording, the disc is turned until the parallel engraved lines on the disc are running parallel to the printed beats of the movement. The gain or loss is then read from the scale on the edge of the disc, and an appropriate adjustment is made to the movement.

In addition to time-keeping accuracy the machine is capable of recording a range of faults, for example, the presence of magnetism, damaged balance staff pivots, faulty escape wheel, faulty fourth

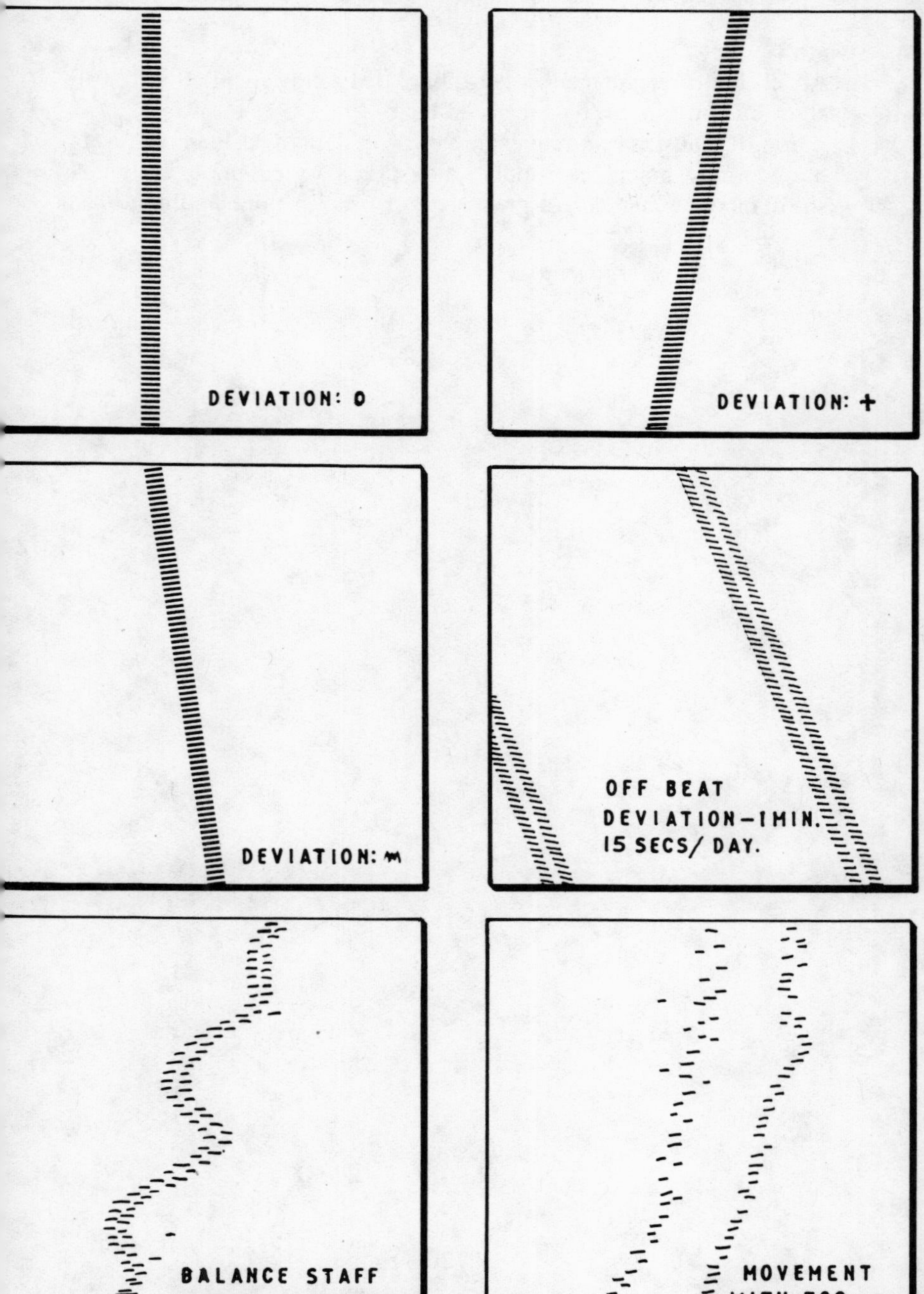

Fig. 85 Examples of traces

wheel, faulty escapement, over-oiling, dirty movement. It will also show by how much a balance is off beat.

Fig. 85 illustrates a few examples of traces. Most manufacturers of recording machines supply trace prints of common faults. A study of these can save a great deal of time in fault finding.

VIII

Calendar Work

THE PRINCIPLE ON WHICH CALENDAR WORK OPERATES IS THE same for clocks as it is for watches, and the only differences are in design detail.

There are two basic types of calendar work; the 'simple' and the 'perpetual.' The simple calendar automatically changes the day of the week and the date (day of the month) every 24 hours. At the end of 31 days the month is automatically changed and the date starts at 1 again. When a month has less than 31 days the date and the month have to be reset by hand.

The perpetual calendar, as the name implies, needs no resetting and, providing the movement is kept wound, it will correctly indicate the day, the date, and the month at any time during any year. The calendar work is designed to anticipate February with its 28 days, the four months with 30 days, and leap years when February has 29 days. The Gregorian reform introduced the concept that centenary years would be leap years only if the number was divisible by 400. This meant that in 1900 all perpetual calendar work was wrong and had to be reset, but the same thing will not happen again until the year 2100.

SIMPLE CALENDAR: (Fig. 86) It is usual for the train to begin with a wheel that is a friction fit on the hour wheel pipe.

This wheel gears into two intermediate wheels that have twice as many teeth, and therefore rotate once to every two revolutions of the hour wheel. These two intermediate wheels are positioned one each side of the hour wheel, and each carries an index pin.

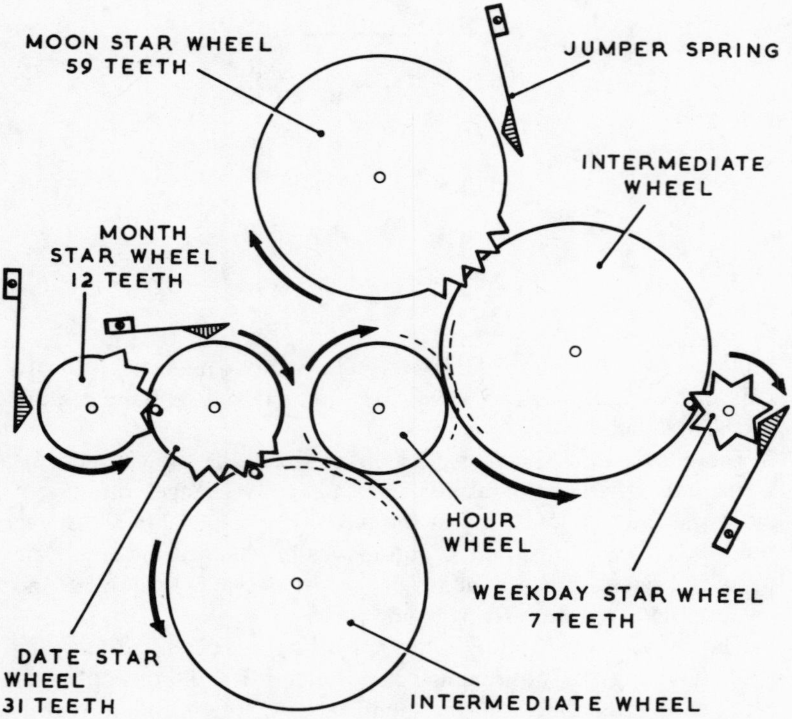

Fig. 86 Simple calendar

Two star wheels are planted in the path of these two pins in such positions as to cause them to move forward one tooth every 24 hours.

One of the star wheels has 7 teeth; this is the day of the week wheel. Two methods can be used to indicate the day and the date: sometimes a hand is fitted to the arbor of this wheel showing the day on a clock dial; in other cases, the arbor can carry a disc on which are painted the days of the week which are visible, one at a time, through a window in a watch dial.

The other star wheel is the date wheel and has 31 teeth. Each time the pin on the intermediate wheel completes a revolution, the date wheel is moved forward one tooth. Here again the date wheel arbor can carry either a hand or a painted disc depending on whether it is a clock or watch.

The date wheel carries an index pin which engages a star wheel with 12 teeth; this is the month wheel. It takes 31 days for the date wheel to complete one revolution and it therefore follows that the index pin will engage the month wheel and move it forward one tooth every 31 days.

All three star wheels are held in their respective positions by jumper springs. When the star wheels are being turned the new tooth pushes against the V piece at the end of the spring and raises it to pass underneath. Immediately after the point of the star wheel tooth passes the apex of the jumper, the tension of the spring pulls the star wheel into its new position and holds it.

Some grandfather clocks are fitted with a moon disc above the main dial. Fitted to one of the intermediate wheels is another index pin which engages a star wheel that is cut with 59 teeth. To this star wheel is fitted a moon disc that is painted with two moons diametrically opposite one another. These moons appear one at a time above the clock dial. The new moon rises from behind a pictorial representation of one side of the world until it reaches full moon size and then begins to drop behind the other side of the world to complete a lunar cycle. Shortly after the old moon is lost to view the other moon begins to appear and a new cycle begins.

One lunar month is equivalent to approximately 29½ days and to enable the moon disc to complete one revolution in two lunar months the star wheel must be cut with $29\frac{1}{2} \times 2 = 59$ teeth.

Some calendar watches are fitted with date change only. Figs. 87 and 88 illustrate the usual arrangement. The hour wheel is in mesh with an intermediate wheel, which in turn is in mesh with a date wheel. The gear ratios are such that the date wheel rotates once in 24 hours.

In the face of the date wheel is a small pin which engages with the 31 teeth on the inner edge of a date ring. For each 24 revolu-

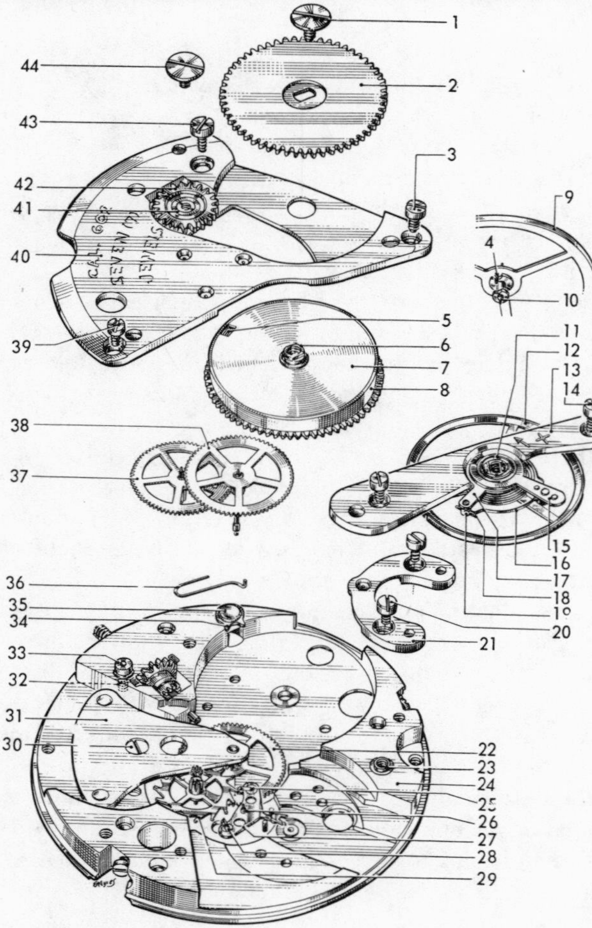

Fig. 87 Lever movement with date change

1 Ratchet wheel screw
2 Ratchet wheel
3 Bridge screw
4 Impulse pin
5 Mainspring
6 Barrel arbor
7 Barrel cover
8 Barrel drum
9 Balance wheel
10 Balance staff
11 Shock-resistant unit
12 Balance wheel
13 Balance bridge
14 Bridge screw
15 Regulator
16 Balance spring
17 Stud arm
18 Balance stud pin
37 Third wheel and pinion
38 Fourth wheel and pinion
39 Bridge screw
40 Train bridge
41 Crown wheel ring
42 Crown wheel
43 Click spring screw
44 Crown wheel screw
45 Locking pad
46 Locking pad spring
47 Setting wheel
48 Sliding pinion
49 Winding pinion
50 Winding stem
51 Setting lever
52 Yoke
53 Yoke spring
54 Retainer screw

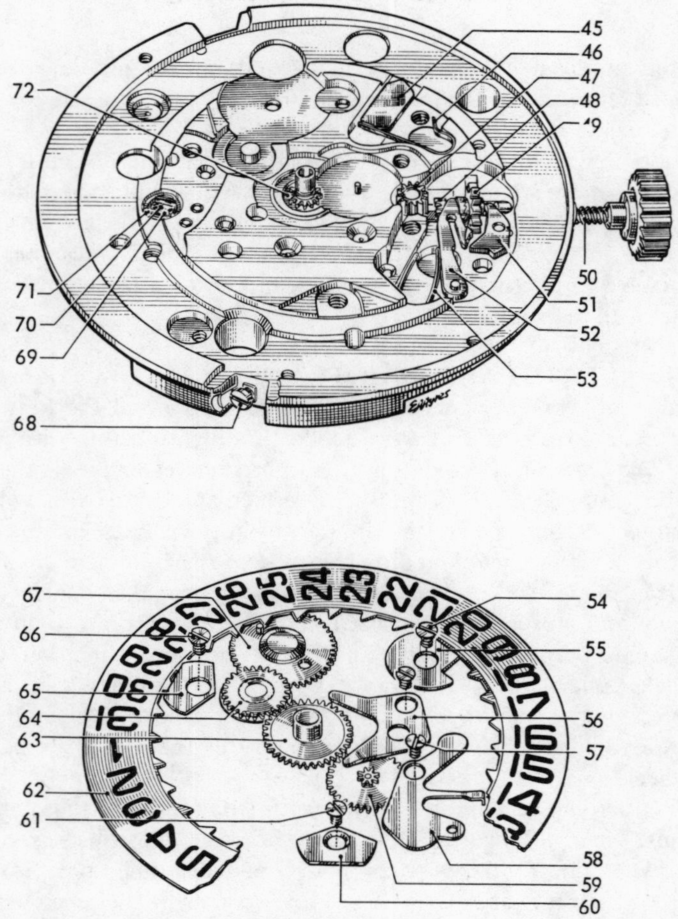

Fig. 88 Lever movement with date change

19 Balance stud
20 Bridge screw
21 Pallet bridge
22 Center wheel and pinion
23 Endshake pillar
24 Plate
25 Pallet stone (entry)
26 Pallet staff
27 Pallet (lever)
28 Pallet stone (exit)
29 Escape wheel and pinion
30 Bridge screw
31 Center bridge and bush
32 Setting lever spring
33 Setting lever post
34 Click
35 Click stud
36 Click spring

55 Date ring retainer
56 Minute wheel retainer
57 Retainer screw
58 Yoke retainer
59 Minute wheel and pinion
60 Date ring retainer
61 Retainer screw
62 Date ring
63 Hour wheel
64 Date intermediate wheel
65 Date ring retainer
66 Retainer screw
67 Date wheel
68 Dial foot screw
69 Shock spring
70 Endstone
71 Ring mount and bearing
72 Cannon pinion

tions of the hour wheel, the date wheel moves the date ring forward one day and the new number is read through a window cut in the dial.

Some watch crystals are made with a small area of optic magnification immediately above the window to enable the date to be more clearly seen.

For months with less than 31 days the hands must be turned forward on the first day of the new month to bring number one in line with the window.

PERPETUAL CALENDAR (Fig. 89) The layout of perpetual calendar work in watches varies considerably and the best advice one can have is to study the action thoroughly before dismantling.

The basic principle of almost all perpetual calendars is the same. The hour wheel gears into an intermediate wheel with twice the number of teeth, causing the intermediate wheel to rotate once in every 24 hours. Mounted on the pipe of the intermediate wheel is a cam and once in every 24 hours the cam makes contact with a lever and moves it to one side against the tension of a light spring.

Some perpetual calendars are fitted with levers that carry three clicks; one is to operate the day of the week star wheel, another operates the star wheel for the moon disc, and the third click acts on a snail which is screwed to the day of the month or date star wheel.

Other watches are fitted with intermediate wheels that carry two pins. One pin engages the day of the week star wheel, moving it forward one tooth every 24 hours, and the other pin moves the moon disc forward. With such an arrangement the lever will need only one click, which is to engage the snail on the date star wheel.

The toe of the lever is so shaped that when the intermediate wheel cam comes into operation and moves the lever to one side the toe of the lever engages the date star wheel and moves it forward one tooth. This happens with each 24 hour revolution of the intermediate wheel, and because the date star wheel is cut with 31 teeth it follows that after 31 revolutions of the intermediate wheel the date star wheel completes one revolution.

This presents no problem when displaying the date of a 31 day month but when the month has fewer than 31 days there has to

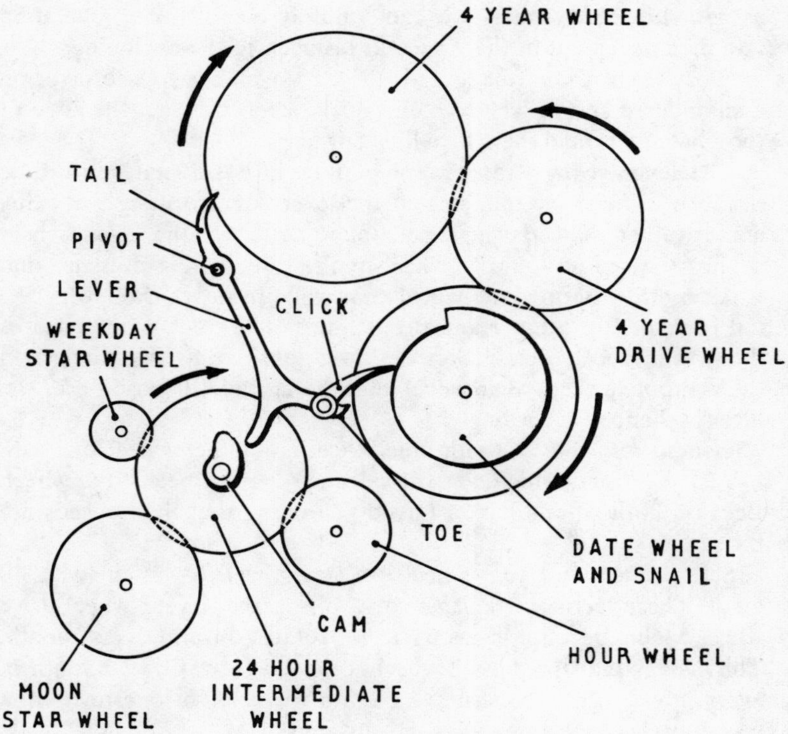

Fig. 89 Perpetual calendar

be an arrangement which will move the date star wheel forward more than 1 tooth at the end of the month. For example, on the last day of a 30 day month the date star wheel needs to be moved forward 2 teeth if the watch is to display the figure 1 on the first day of the new month. On the last day of 28 days in February the date star wheel must be moved forward 4 teeth, and during a leap year the date star wheel must be moved forward by 3 teeth on the last day of February. It is this requirement that makes the difference between a simple calendar and a perpetual calendar.

Screwed to the date star wheel is a steel snail. This is a flat plate shaped like the creature from which it derives its name. It functions as a continuous progressive cam until it completes a revolution and then a step cut in the edge takes it back to the beginning.

Close to the toe of the lever is the click which is free to pivot on a shouldered screw. The free end of the click rides on the edge of the snail and is held there by a light spring.

The lever swings about a pivot that is positioned some little distance in from the tail, so that the lever is performing a rocking action rather than a swinging action. The tail of the lever is held against the rim of a 4 year wheel by the same spring pressure that is felt by the intermediate wheel cam when it moves the lever. The tail is shaped to a taper and drops into notches cut in the edge of the 4 year wheel. This makes the lever sensitive to any changes in the contour of the 4 year wheel, and the attitude or position of the lever is changed accordingly.

Geared into the 31 tooth date wheel is another wheel that also has 31 teeth and which we will call the 4 year wheel drive wheel. Because both wheels have 31 teeth it follows that their speeds are the same.

Screwed to the 4 year wheel is a star wheel that is cut with 48 teeth. These teeth are engaged by a pin in the 4 year wheel drive wheel which, because it has 31 teeth, rotates once in every month. This means that the 4 year wheel is moved forward by 1 tooth in every month. Because of its 48 teeth it follows that one revolution takes 4 years to accomplish; hence its name.

These are the parts that make up the perpetual work, so let us take a closer look at their synchronized functions. Fig. 90 illustrates diagrammatically a 4 year wheel showing the notches cut in the edge. This is a much exaggerated drawing for the purpose of clarity. During a 31 day month the tail of the lever rests on the full diameter of the wheel. A 30 day month is represented by a shallow notch cut in the edge of the wheel. A leap year February of 29 days has a deeper notch, and 28 day Februarys have full depth notches. The illustration shows only 14 months but the wheel is cut all the way round and all four quarters are identical with the exception of February in the leap year.

During any 31 day month, with the tail of the lever on the maximum diameter, the intermediate wheel completes 31 revolutions and with each one the cam pushes the lever over and causes the toe to engage the date star wheel and move it forward 1 tooth. At the end of the month the date star wheel will have completed 1 revolution.

We have already seen that screwed to the date star wheel is a snail on which the click is riding. As the days of the month pass and the date star wheel moves round so the step in the snail gets closer to the click until, on the day before the last when the toe of the lever moves across to engage the date star wheel, the click also engages the step in the snail and both click and toe advance the date star wheel by 1 tooth.

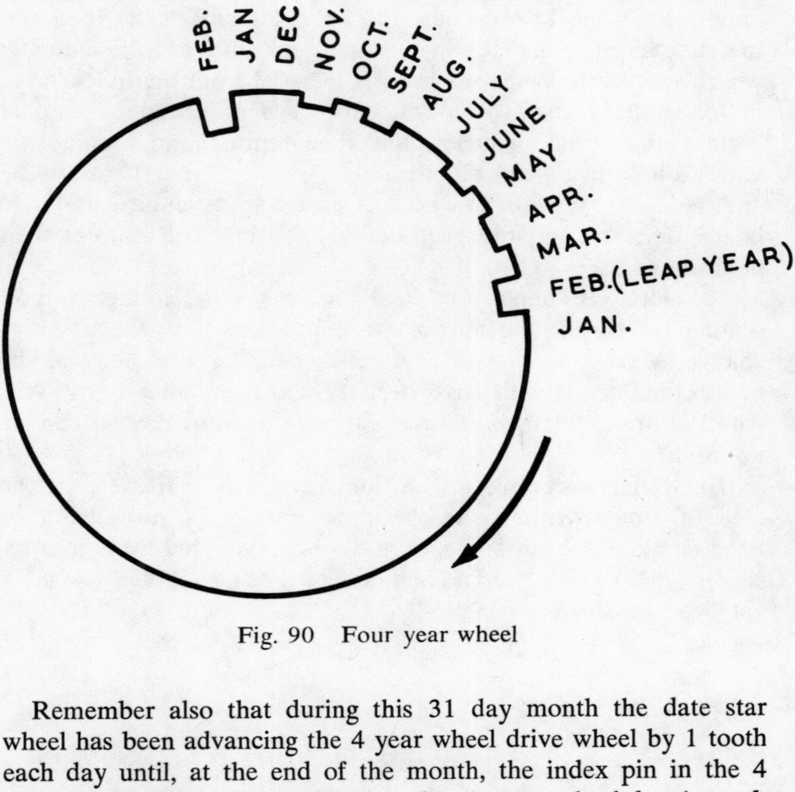

Fig. 90 Four year wheel

Remember also that during this 31 day month the date star wheel has been advancing the 4 year wheel drive wheel by 1 tooth each day until, at the end of the month, the index pin in the 4 year wheel drive wheel advances the 4 year wheel by 1 tooth which is equivalent to 1 month. While the 4 year wheel is being advanced the cam of the intermediate wheel is engaging the lever which holds the tail of the lever clear of the 4 year wheel. When the 4 year wheel has completed its forward movement the intermediate wheel cam releases the lever allowing the tail to return to the notched contour. If the completed month had been January, the tail of the lever would be resting in the deep notch of February.

In this new position the toe of the lever is taken further from the date star wheel allowing the click to slide round the edge of the snail and taking up a position closer to the step. The previous cycle is repeated but on the completion of 28 days the click, in its new position in relation to the snail, engages the step and when the intermediate wheel cam comes into operation and moves the lever, the click engages the step in the snail and advances the date star wheel by 4 teeth. When the lever is released from the influence of the cam, the 4 year wheel will have been advanced by another tooth and the full diameter, this time representing March, will once again be in contact with the tail.

After calendar work has been cleaned and assembled it should be left dry. The only oiling needed is to springs and cams at their points of contact.

The setting is simple. The hands are put to 12 o'clock (representing midnight) and all parts are positioned on the point of change. Make sure that the correct day of the week, day of the month, and the month have been selected, and that the 4 year wheel has been fitted so that the leap year is in the correct relative position.

If calendar work fails to function correctly due to wear on the pieces the only remedy is to obtain new pieces. Do not attempt to bend or file any of the parts because this invariably leads into reshaping other pieces, and a good deal of time can be spent without achieving satisfactory results.

Part Two

WATCHES

IX

Servicing

THE MAJORITY OF WATCHES THAT NEED THE SERVICES OF A RE-pairer do so for one of three reasons:

1. The watch functions but does not maintain correct time.
2. The watch has stopped and cannot be restarted.
3. The customer requires that his watch be given a periodic clean.

In the first instance the gain or loss can be steady or erratic. If it is steady and the movement appears to be clean and adequately lubricated, it is more than likely that adjustment will put the matter right.

If the gain or loss is erratic then one must look for the cause before the fault can be rectified. A spot of oil on the hairspring will set up an irregular gain, a loose hand will register fast time on one occasion and may indicate slow time on another occasion. A hand binding against the crystal in one spot can cause an erratic loss, while the pipe of a small second hand occasionally touching the hole in the dial will cause the watch to stop and the movement of the wearer's wrist will restart the watch again.

If the watch stopped because of some reason other than the lack of winding, it is most likely to be the fault of a broken balance staff, a dirty movement, or a broken mainspring.

Some people like to have their watches cleaned periodically and the repairer is expected to correct anything that is found to be in need of attention. Manufacturers of fine watches recommend regular and frequent cleaning. For instance, Longines recommends that their watches be overhauled every twelve to eighteen months for small movements, and every two years for larger models.

When a watch has been accepted for repair it is important that it be closely examined before anything is disturbed because of the possibility of discovering a defect that dismantling might hide.

The appearance or general condition of the outside of a watch often indicates the type of conditions in which the watch has had to function, and can sometimes be helpful in identifying the cause of the breakdown.

If the watch is not going, make sure that it is wound up. Oscillate the watch to set the balance in motion and then apply extra power to the gear train by exerting a small amount of pressure on the winding crown. If the movement responds immediately and gives audible evidence of functioning, and then stops when the hold on the winding crown is released it is almost certain that the trouble is caused by dirt and lack of lubrication.

A center sweep hand can catch on the dial or another hand or on the crystal. Any one of these can stop a movement. Examine the hour and minute hands to ensure they are not catching each other. Pull out the winding stem into the hand set position and slowly rotate the hands and at the same time hold the watch to your ear. If the hands are fouling the dial or the crystal, a scraping or scratching sound will be heard.

Look at the back of the watch case to see if it has been dented. With some watches there is very little clearance between the movement and the case and it requires very little pressure on the balance cock to take up the balance staff end float and so stop further vibrations. Some gold cases have very thin case backs and if the wearer overtightens the wristband or bracelet the watch case is pulled against the wrist hard enough to cause the case back to press on the movement.

Watch repairers are usually busy in the season when vacationers return from their trips. Salt water and sand take a heavy annual toll of watches.

The percentage of watches that are brought in with broken mainsprings is lessening due to modern manufacturing techniques, but even so there are still plenty of mainsprings renewed each year.

The fitting of shock resistant bearings for balance staff pivots has been by far the greatest contribution towards the prevention of broken pivots, and in the majority of cases when a watch has been subjected to shock the pivots have remained unharmed. There are circumstances when the pivots are subjected to side forces too great to be absorbed by the units and the pivots shear. This frequently happens if a watch is dropped onto a concrete floor.

When a watch is brought to you for attention it should be received in businesslike manner, and carefully recorded so that there is no doubt as to the identity of the watch owner. This will promote confidence in the customer who will be satisfied that the watch is in good hands.

All that is required is a duplicate order book that has the pages printed numerically, and some tie-on labels. At the top of the duplicate order book write the customer's name, address, telephone number, and date. Then underneath a very brief description of the watch with any details supplied by the customer that may be helpful during the repair. On one of the labels write the printed order number and the customer's name and tie the label to the wristband. Tear out the carbon copy of the order and hand it to the customer as his receipt.

The printed order number now becomes the repair number and it should be scratched inside the back of the case for future reference. It will provide positive identification if the watch is subsequently brought to you again for other work. The number should be preceded by your initials to prevent confusion with other repairers, e.g., HH 32.

A suitable tool for this purpose is an engineer's scriber that has been sharpened on an oilstone. The numbering should be small and carried out with the aid of a loupe. It is best done inside the back of the case close to the edge as neatly and inconspicuously as possible.

In the workshop keep a register of all repairs and enter them in numerical order. Suggested column headings are shown below but these can be altered to suit individual requirements.

REPAIR NO.	DATE	CUSTOMER'S NAME	TYPE OF WATCH	WORK DONE AND MATERIAL SUPPLIED	COST

The sequence of disassembling a watch must vary according to its layout and its function. Examples of extra mechanisms are chronographs, repeaters, simple and perpetual calendars, and alarm and musical watches, but these are additional to the ordinary movement, which is common to them all.

Fig. 91 shows an exploded view of an English 17 jewel 12 ligne center second movement fitted with stop work with all parts annotated and identified in the index. It will serve as a useful guide for future reference.

After our preliminary external examination, the next step is to remove the movement. If the watch is a water resistant model with a sealed case, reference should be made to Chapter XII for the correct method of opening.

If the back of the case is a snap-on type, it is recommended that a case knife be used. This is a knife with a single short fixed blade shaped especially for this operation. The edge of the blade is carefully inserted between the back and the body of the case, and then rolled over on its side. If the blade has been inserted to the full depth of the recess, the back will spring off. Some case manufacturers make a small curved cut-away portion to facilitate the entry of the knife blade.

Never twist the knife. This can only lead to damage to the case, and in some instances permanent disfigurement, particularly if the case is made of gold. The knife cuts into the edge of the back and the edge of the case body and throws up burrs. Further twisting usually causes the blade to ride up and out of the recess and if it slips on the case it will leave behind a skid mark.

The same procedure is applied to the removal of snap-on crystal bezels. Sometimes when a case back or crystal bezel is removed, the movement starts to go when previously nothing would induce it to function. This is a sure sign that when the case was assembled either the case back was pressing against the movement or the crystal was touching a hand.

The presence of dust will indicate a bad fitting case or crystal. If the crystal is at fault the matter can usually be put right by running a very small quantity of crystal cement round the edge with an old clock oiler. Shaped cases, other than round ones, are frequently troublesome, particularly the two piece cases where the bezel fits over a flange of the back. A smear of beeswax and vaseline, as mentioned in Chapter XII usually produces an effective seal. It can also be used in winding stem pipes which, in some low priced watches, offers an open invitation to the ingress of dirt and dust.

The presence of rust on the winding and hand set mechanisms is an indication that water has passed through the winding aperture, and is more often than not the result of the wearer placing his or her hands in water without first removing the watch. If the corrosion is only surface rust it can be removed by careful filing but if it is deep and the metal is pitted new parts are advisable.

When the halves of a two piece case have been separated, the movement can be lifted out of the back, but with a three piece case the winding stem and case screws must first be removed before the movement will drop out. If the pull-out piece screw is slackened, the pull-out piece will become disengaged from the winding stem making it possible to withdraw the winding stem from the movement.

It is safer and easier to remove the hands while the movement is still in its case. The hands are lifted by means of a pair of watch hand removing levers supplied for the purpose. Folded pieces of tissue paper are placed beneath the levers to prevent damage to

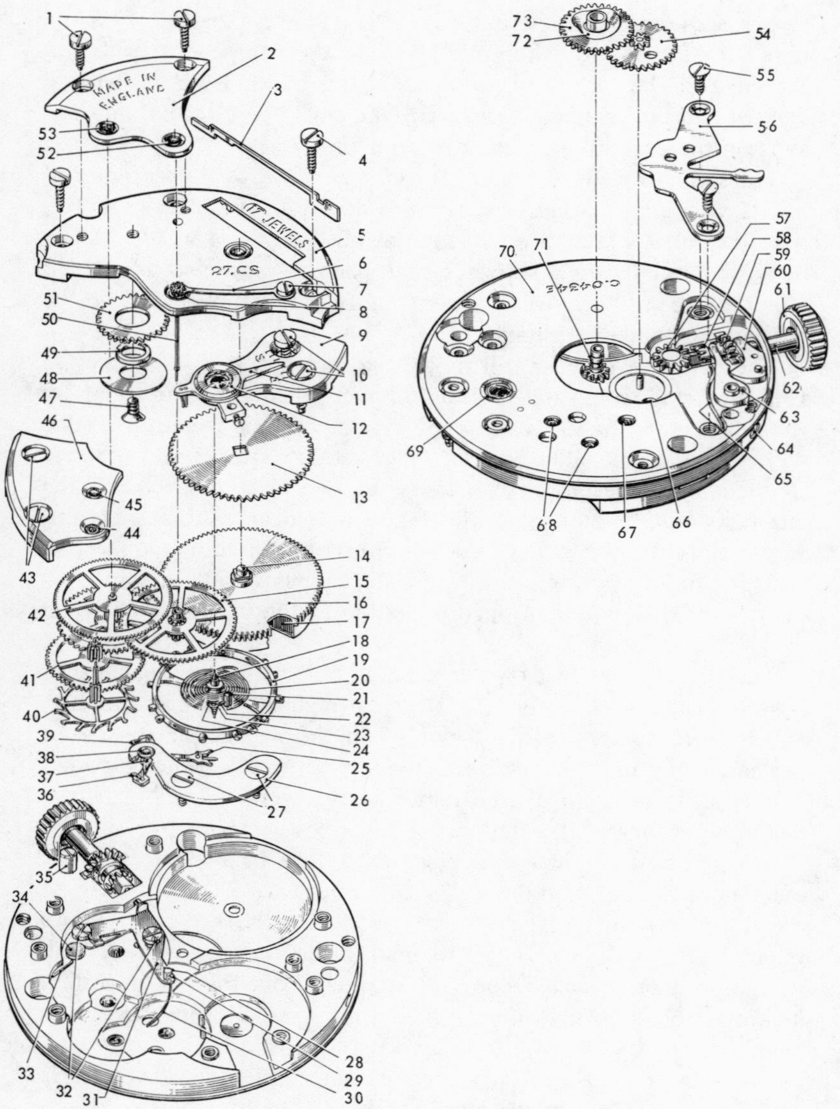

91 English 17 jewel 12 ligne center secor

1 Center bridge screws
2 Center bridge
3 Click spring
4 Bridge screws
5 Barrel bridge or top plate
6 Friction spring
7 Friction spring screws
8 Center jewel
9 Balance cock
10 Balance cock screws
11 Regulator
12 Upper shock-resist unit
13 Ratchet wheel
14 Barrel arbor
15 Center wheel
16 Barrel and lid
17 Mainspring
18 Balance staff
19 Balance
20 Hairspring
21 Hairspring stud
22 Hairspring pin
23 Roller
24 Ruby pin
25 Escape lever or pallet lever
26 Lever bridge or pallet cock
27 Lever bridge screws
28 Stop lever blade stud
29 Stop lever pin
30 Spring blade
31 Stop lever
32 Stop lever screws
33 Stop lever spring
34 Stop lever spring stud
35 Pull-up piece screw
36 Pallet stone (exit)
37 Pallet staff
38 Escape, pallet and center spindle jewels
39 Pallet stone (entry)
40 Escape wheel
41 Fourth wheel
42 Third wheel
43 Bridge screws
44 Escape, pallet and center spindle jewels
45 Third jewels (plate and bridge) and fourth jewel (bridge)
46 Wheels bridge
47 Crown wheel screw
48 Crown wheel retaining plate
49 Crown wheel sleeve
50 Center spindle
51 Crown wheel
52 Escape, pallet and center spindle jewels
53 Third jewels (plate and bridge) and fourth jewel (bridge)
54 Minute wheel
55 Setting lever spring screws
56 Setting lever spring
57 Setting pinion
58 Sliding pinion (castle wheel)
59 Winding pinion
60 Pull-out piece
61 Winding crown
62 Winding stem
63 Return lever
64 Dial screws
65 Return lever spring
66 Third jewels (plate and bridge) and fourth jewel (bridge)
67 Fourth jewel (plate)
68 Escape, pallet and center spindle jewels
69 Lower shock-resist unit
70 Bottom plate
71 Cannon pinion
72 Dial washer
73 Hour wheel

the dial. If the watch is fitted with a small second hand this is best withdrawn by the dial, when the dial is removed.

Before taking the movement from its case it is necessary to let down the mainspring. Turn the winding crown very slightly until the click is lifted by the next tooth on the rachet wheel. With a pointed pegwood stick hold the click clear from the influence of the ratchet wheel, and slowly allow the winding crown to rotate between finger and thumb until the mainspring is fully unwound. The click can then be released.

Lift the movement from the back of a two piece case by easing it out with the winding stem. If the fit is tight, place the blade of a small screwdriver beneath the bottom plate and gently ease the movement out. Take care not to make the mistake of inserting the blade beneath the dial. Leverage under these conditions would cause permanent disfigurement of the dial, which would then have to be replaced. This would have to be at the expense of the repairer and dials are sometimes expensive. There is also the very real risk that an identical dial may not be available.

With a three piece case it is usual for the movement to be held in position by two case screws. These screws are threaded into the underside of the bottom plate and are positioned so that the screw heads protrude over the edge of the plate and behind the lip of the case center piece. In proportion to the diameter of the thread, these screws have large heads. Some of them are completely round, while others have had a flat cut in the edge.

The completely round screws must be removed before the movement will drop out of the case. The flat-sided screws need to be turned only until the flat coincides with the edge of the bottom plate. In this position the screws have no influence over the security of the movement, which will simply drop out.

The dial is next to be removed. It is common practice today to insert two small screws into the side of the movement in line with the two copper feet of the dial. When the dial has been placed in position, the two screws are turned until they bite into the soft metal of the feet and the dial is held. An older method is to fit two dog screws into the bottom plate. These screws are made with a conical flange halfway along their length. A half-moon piece is cut from the edge of the cone to provide clearance for the dial feet.

The screws are inserted into the underside of the bottom plate, and screwed in as far as they will go. They are then turned counter clockwise until the half-moons coincide with the dial feet. The dial will then drop down into its correct position and the two screws can be turned in a clockwise direction to cause the edge of the conical flange to bite into the copper feet.

The dial is raised by careful levering with the blade of a knife placed close to one foot and then the other. Keep the dial as level as possible and take care not to lose the small second hand if one is fitted.

When the dial has been removed, note whether or not a dial washer had been fitted over the pipe of the hour wheel. These washers are made from brass foil and are slightly curved to give them spring tension. They are used to prevent the hour wheel from riding up the cannon pinion and causing the hour hand to foul the minute hand. These washers are sometimes lost and forgotten and dials are refitted without them.

Next to be removed is the motion work. Lift the hour wheel from the cannon pinion and the minute wheel from the minute wheel post, and place them in a covered parts tray, Fig. 13. These trays protect the parts from loss and damage and a repairer should develop the habit of placing every part in a tray immediately after it has been removed from the watch.

Select a movement rest, Fig. 14, that will support the bottom plate, and place the movement on the rest with the top plate uppermost. Inspect the hairspring to see if there is any obvious damage or misalignment. The spring should be quite central and horizontal. If the spring is seen to be coiled to one side, or out of horizontal with the rest of the movement, one can suspect that the damage has been caused by inexperienced fingers and a pin. If the hairspring is in need of attention this will have to be done when the balance assembly has been removed and disassembled.

Examine the index pins to see what type has been fitted. If one index pin and a turnboot is in use, Fig. 92 (a), the balance cock, complete with balance and hairspring, must be removed as an assembly. If there are two index pins, Fig. 92 (b), then the balance cock can be removed separately.

Turnboots cover the bottom of the index pin, and are used to prevent other coils of the hairspring jumping in when the watch is subjected to shock. Watches with flat hairsprings operate with either two index pins or one pin with a turnboot.

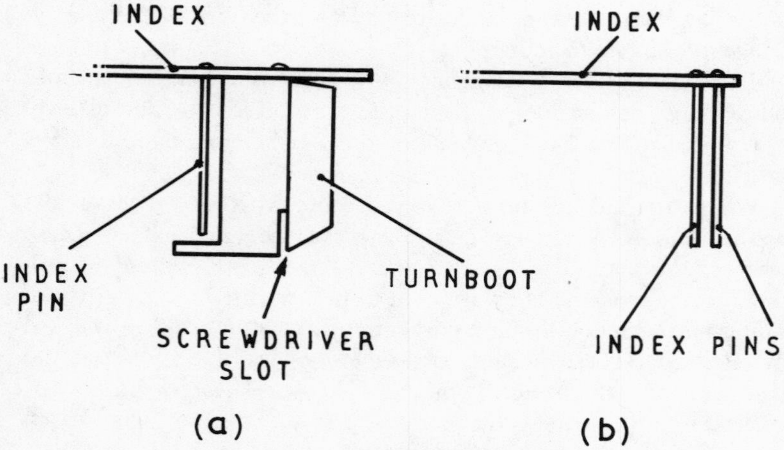

Fig. 92 Index pins

If a watch is fitted with a hairspring that has an overcoil, there is no need for a turnboot because it is very unlikely that one of the coils would jump into the index.

In the case of a turnboot, remove the balance cock screw and gently pry the balance cock locating pins from the bottom plate by inserting a small screwdriver blade in the slot of the balance cock foot. When the cock is free, carefully lift it from the movement making sure that the hairspring does not catch any of the wheel teeth. The hairspring must not be stretched because it will be impossible to restore it to its original flat shape.

If the balance shows reluctance to come away from the movement, it will be because the roller on the balance is fouling the pallet lever. If this happens the cock must be held stationary while the movement is turned first one way and then the other. As soon as the roller has freed itself the balance will come away and hang freely on the hairspring.

Gently lower the assembly onto the bench with the cock resting on top of the balance. Carefully roll the cock over onto its back and follow it with the balance. It is advisable not to put the balance staff pivot into the jewel hole in the balance cock, in case some pressure should accidentally be applied to the balance and snap the pivot.

Turnboots are either slotted for a screwdriver or drilled in the side to take a pointed instrument. Either way, the turnboot is rotated until the index pin is clear and the hairspring coil is free to be withdrawn. Slacken the screw in the balance cock and with tweezers lift out the hairspring stud and hairspring.

When two index pins are employed, the hairspring stud screw is slackened, the stud is pushed out, and only the cock is removed from the movement. The balance complete with hairspring is then lifted out separately.

The hairspring should be sitting flat and parallel to the balance and should also be central about the balance staff. If the hairspring has been damaged so that it is out of position, the spring must be corrected by reshaping at the point where it has been bent. Two pairs of hairspring tweezers are needed for this operation. Place the tweezers on each side of the bend or kink, and bend the spring until the correct set has been restored.

Damage to hairsprings occurs most frequently in the vicinity of the hairspring stud, and sometimes the damage includes a slight twist as well as a bend. In such instances, the hairspring must be removed from the balance and laid flat on a sheet of white paper before correction with the tweezers can be carried out. When work on the balance assembly is complete, place all pieces in one of the compartments of a parts tray.

With the balance removed, give the winding crown a few turns to put tension in the mainspring. Then take a pointed pegwood stick and very slowly push against one side of the pallet lever close to the guard pin. The tension in the wheel train should cause the escape wheel teeth to give an impulse to the lever pallets, so that the lever will jump away from the pegwood and rest against a banking pin. Repeat the test in the opposite direction. If the lever does not jump but has to be pushed all the way, it will be necessary to remove the lever to locate the area in which the fault is occurring.

Let down the mainspring and remove the pallet cock and the lever. There is now nothing to prevent the gear train from spinning if the mainspring is wound. Turn the winding crown very slightly and observe the reaction of the escape wheel. If the train is clean and free, the escape wheel will move immediately. When the tension is released, the escape wheel will stop and make a few revolutions in the reverse direction. If this happens there can be nothing wrong with the freedom of the wheel train. The fault would appear to be with the pallet lever and the most likely cause would be dirty and dry pivot holes.

If there is any hesitancy on the part of the escape wheel to rotate at the slightest pressure, then the movement must be disassembled and cleaned.

X

Stop Watches and Chronographs

Stop Watches

THE ORDINARY STOP WATCH, FIGS. 93 AND 94, DOES NOT INDICATE mean time (time of day), but measures the amount of time between any two given moments. Some movements are capable of measuring to one hundredth of a second but the more usual stop watches measure to one tenth or one fifth of a second.

The movement carries a sweep second (center second) hand which indicates the number of seconds on what would be the minute track of a conventional watch. Each revolution of the sweep second hand is recorded by a minute recording hand, that has its own small dial printed on the main dial.

The movement is stopped by depressing the winding crown. The inward movement of the winding stem depresses an actuating arm, which engages a tooth on a split wheel, and causes it to turn. The split wheel has twelve rachet type teeth around its edge and four concentric blocks or pillars machined integrally on its face. By depressing the winding crown three times, and thereby turning the split wheel by the distance of three teeth, the next pillar is brought into line ready for the next cycle.

When the wheel is turned one tooth, one of the pillars is pushed against the end of the stop lever causing it to rock about its fulcrum. At the other end of the stop lever is a piece of fine brass wire that

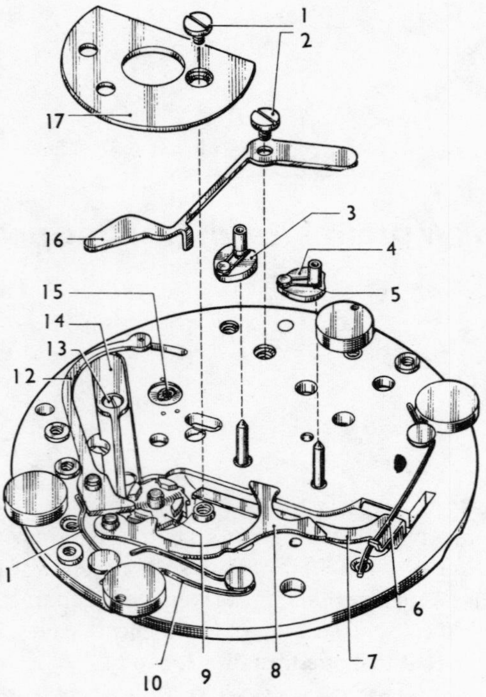

Fig. 93 Stop watch

1 }	Brake and mechanism	10	Striker spring
2 }	bridge screws	11	Stop lever spring
3	Seconds cam	12	Stop lever
4	Minute cam	13	Split wheel screw
5	Cam spring	14	Split wheel spring
6	Actuating arm spring	15	Lower shockproof unit
7	Actuating arm	16	Center wheel brake
8	Striker	17	Mechanism bridge
9	Split wheel		

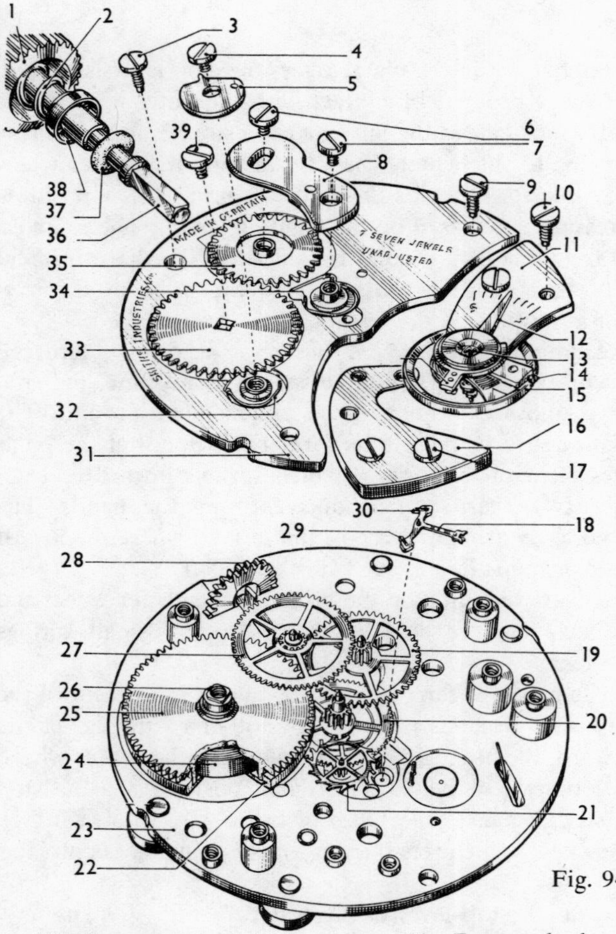

Fig. 94 Stop watch

1	Winding crown	21	Escape wheel
2	Crown spring	22	Barrel lid
3	Bridge screw	23	Plate
4	Winding wheel screw	24	Mainspring
5	Click	25	Barrel
6, 7	Winding wheel screws	26	Barrel arbor
		27	Second wheel
8	Crown wheel rocker	28	Winding pinion
9, 10	Bridge screws	29	Pallet stone (exit)
		30	Pallet stone (entry)
11	Balance cock	31	Three-quarter plate
12	Regulator	32	Click spring
13	Endstone	33	Ratchet wheel
14	Hairspring stud screw	34	Crown wheel rocker spring
15	Balance	35	Crown wheel
16	Escape bridge	36	Winding stem
17	Bridge screws	37	Sealing ring
18	Escape lever	38	Top gland housing
19	Third wheel	39	Click, ratchet, crown wheel and rocker screw
20	Center wheel		

protrudes through a slot in the plate. When the stop lever is moved, the piece of wire is brought in contact with the outer edge of the balance wheel, which causes the movement to stop. Simultaneously, a brake is applied to the center wheel by moving the center wheel brake lever. A side spring holds the split wheel in its new position, even when pressure is released on the winding crown. The minutes, seconds, and fractions of a second recorded on the dial represent the interval of time that has passed from the moment the stop watch was put in motion, to the instant when it was halted.

Once the reading has been taken, the hands need to be returned to zero. By depressing the winding crown a second time, the split wheel is turned another tooth and the pillars move around. The stop lever continues to be held in its former position, but the pillar in front moves clear of the striker, which springs inward and engages with the two heart-shaped cams carrying the hands. The cams are forced to fly around to a position of rest, where the lowest part of the cam remains in contact with the striker.

The two cams are mounted on the pivots of the center wheel and the second wheel, and are held by the friction of small springs carried on the cams.

Having set the hands to zero, a third downward movement of the winding crown will turn the split wheel another tooth, the pillars will move forward, the brake will be released from the center wheel, and the stop lever will move to its original position. As the stop lever moves back, it causes the piece of brass wire to give a flick to the balance wheel, which sets the movement going and completes the cycle.

Another use for a stop lever is illustrated in Fig. 91. This is a conventional sweep second movement with the addition of a stop lever. When the winding crown is pulled outward, a stop lever moves under the tension of a spring, and carries a curved spring blade which arrests the movement of the balance wheel.

With the movement now stationary, the hands can be set and synchronized accurately to any standard or master timepiece.

The action of depressing the winding crown returns the stop lever to its former position, and at the same time causes the spring blade to give a starting impulse to the balance wheel.

The sweep second hand is mounted on a center spindle, which

passes through the center of the center wheel. The spindle is supported at its pinion end by a friction spring, and is driven by the third wheel.

Chronograph

A chronograph is a conventional watch, to which has been added a mechanism that enables intervals of time to be recorded without interfering with the watch movement.

In addition to the ordinary second hand, the watch has a sweep second hand. The complete revolutions of this are recorded by a small hand and dial, usually situated in the 9 o'clock position on the main dial.

The sweep second hand moves around the dial once in sixty seconds and therefore the small recording hand indicates the number of complete minutes. Such a movement is usually referred to as a minute recording chronograph.

The design and layout of chronograph mechanisms are different with each make of movement and frequently vary between models of the same make. It is therefore better that the basic principles be understood, and then the function of any particular layout will become clear after a few minutes of study. The mechanism is mounted above the top plate of the movement and is exposed by removing the back of the case.

A refinement to the minute recording chronograph is the addition of a split second hand, which is fitted either above or below the sweep second hand, and has its own controlling push piece. This additional hand can be stopped whenever desired, and is independent of the sweep second hand. A typical application of the two hands would be to use the split second hand to time one lap of a race while the sweep second hand recorded the total time of the race.

When the reading indicated by the split second hand has been noted the hand can be released, and it will rejoin and rotate with the sweep second hand.

Control of the sweep second hand is effected by depressing a plunger in the winding crown, or in some watches a separate push piece is provided. The first press will set the mechanism in motion, the second press will stop the mechanism so that a reading may

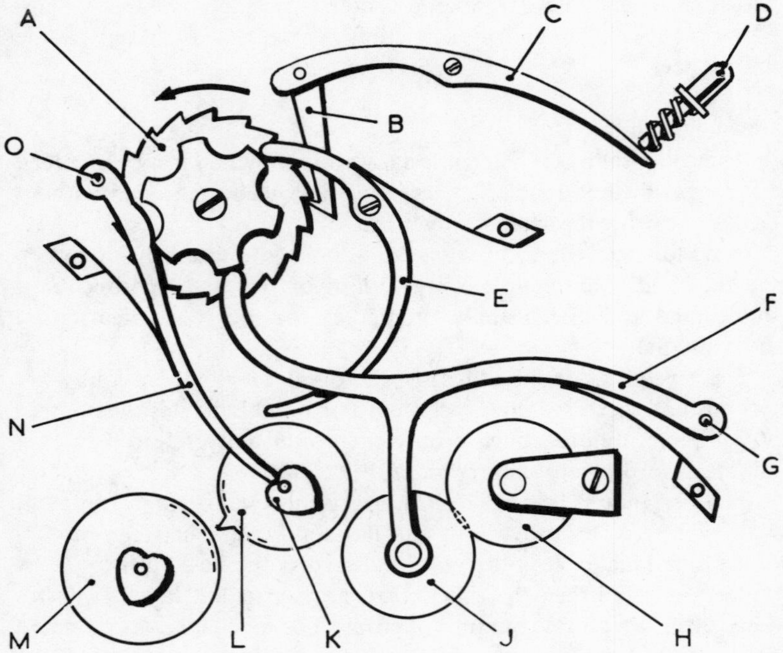

Fig. 95 Simple chronograph

A	Six post crown or ratchet wheel	H	Driving wheel
B	Pawl	J	Transmission wheel
C	Operating lever	K	Heart shaped cam
D	Push piece	L	Center wheel
E	Brake lever	M	Minute recording wheel
F	Transmission lever	N	Zero or fly back lever
G	Pivot	O	Pivot

be observed, and the third press will return the mechanism to zero which means that the sweep second hand and the minute recording hand will return to the 12 o'clock position. There they will remain until the mechanism is set in motion again by a fourth press.

All the time that the chronograph mechanism is in use, the going train is unaffected and with a correctly adjusted movement there will be no gain or loss in the timekeeping.

Fig. 95 shows a diagrammatic layout of a simple chronograph with the mechanism in the zero position. To include all the pieces would be to produce a detailed drawing that may prove to be a little confusing. Our purpose is to aquire an understanding of the basic principles upon which the mechanism is allowed to function, and so only those parts essential for this purpose have been drawn.

The ratchet wheel A has six posts, all of which, in their turn, control the positions of levers E, F and N. The ratchet wheel also has eighteen teeth which means that for every pillar there are three teeth. This is because there are three functions of the push piece to complete one cycle, i.e., start, stop, and return to zero. Some chronographs are fitted with ratchet wheels cut with twenty-four teeth and some with sixteen teeth which have eight pillars and four pillars respectively, but the majority of present day chronographs have ratchet wheels with six pillars and eighteen teeth.

The two wheels H and L are identical in size, and are each cut with the same high number of very fine teeth, usually about three hundred. Wheel J is cut with fine teeth to the same pitch but the wheel is not necessarily the same diameter.

The fourth wheel pinion of the going train of the watch has a long pivot at each end. The ordinary second hand is fitted on the front pivot in the conventional way, but the back pivot is used to carry wheel H which will rotate once in every sixty seconds. This, then, is the beginning of the wheel train for the chronograph mechanism.

Wheel J is carried on an arm of lever F but regardless of the position of lever F, wheel J is always geared to wheel H and so the two wheels rotate together.

In the zero position, as drawn, wheel J is not geared to wheel L but if lever F were allowed to swing in a clockwise direction about pivot G, then wheel J would roll round the circumference of wheel H until the three wheels were all in gear.

When push piece D is depressed, the operating lever C pivots on a shouldered screw and lifts pawl B. The pawl is held against the ratchet wheel by a light spring and when the pawl is raised it moves the ratchet wheel round by one tooth. When the push piece is released, the operating lever returns to its former position. The pawl rides over the following tooth and hooks itself, under spring tension, behind the tooth in readiness for the next sequence of movement.

The action of turning the ratchet wheel causes one of the pillars to move lever N about pivot O against the tension of a light spring. The opposite end of lever N is lifted clear of the heart piece K, and because the heart piece is fixed to wheel L, the wheel is now free to rotate. Similarly, a fourth lever, which is not shown in the drawing, is lifted clear of the heart piece on wheel M.

At the same time that this is happening, another pillar, which has been holding lever F against spring tension, is moved away from the lever. The spring causes lever F to swing about pivot G and the end of the lever is pushed into the crescent behind the retreating pillar. In so doing wheel J is brought into gear with wheel L and the three wheels rotate together.

It follows that because wheel H rotates once in every sixty seconds wheel L, which is identical, must do the same. The arbor that carries wheel L passes through the hollow center pinion of the watch and on the front end is fitted the sweep second hand.

The chronograph mechanism is now in motion. The sweep second hand, which is mounted on the arbor of wheel L, is rotating once in every sixty seconds and will continue until push piece D is again depressed.

Secured to the underside of wheel L is a finger piece which engages the minute recording wheel M. Each time wheel L completes a revolution, the finger piece turns wheel M one tooth. The minute recording wheel is mounted on an arbor, which passes through to the front of the watch. On the front end of the arbor is a hand that rotates above a small dial and indicates the interval of time, in complete minutes, that has passed since the chronograph was set in motion. Some chronographs are required to record only up to thirty minutes. In these cases the minute recording wheels are cut with thirty teeth and the small dial has thirty graduations. Other chronographs which are required to record up to sixty minutes are fitted with a minute recording wheel with sixty teeth and have a dial suitably graduated. The watches are respectively known as thirty minute and sixty minute recording chronographs.

A jumper spring is used to hold the minute recording wheel steady. This is necessary to ensure that the hand points exactly to a minute division on the dial each time it is moved.

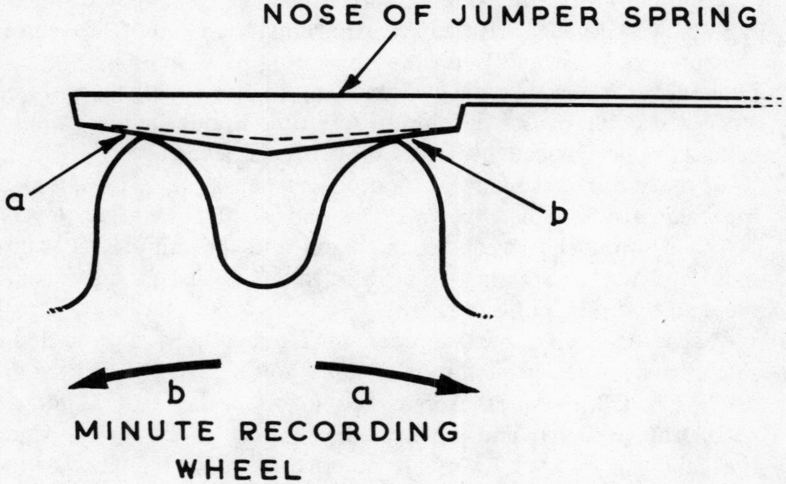

Fig. 96 Adjusting a jumper spring

Another function of the jumper spring is to steady the wheel in order to allow the finger piece to make engagement. If the jumper spring is incorrectly adjusted or positioned, and the minute recording wheel is allowed some free movement, it is possible for the finger piece to foul a tooth and stop the watch.

There are some chronographs that do not use a finger piece to drive the minute recording wheel. The drive is accomplished by gearing the minute recording wheel to the train. Here the minute recording hand is rotating slowly all the time the sweep second hand is moving.

Depressing the push piece a second time and turning the ratchet wheel another tooth will cause the mechanism to stop. Lever F will be moved over to its former position by the advancing pillar, thus taking wheel J out of gear with wheel L. At the same time the end of lever E will drop into a crescent between two pillars on the ratchet wheel, causing the opposite end of the lever to touch wheel L and act as a brake.

Operating the push piece a third time and advancing the ratchet wheel another tooth, will return the hands to zero by what is known as the fly-back action.

Secured to the center of a wheel is a steel, heart-shaped cam with highly polished edges. Bearing hard against the edge of the cam is the end of a lever held in position by spring pressure. The pressure of the lever against the edge of the cam, together with the reduced surface friction, causes the cam to spin round until the lowest point of its periphery is caught by the probe of the lever.

In the simple chronograph that we are considering, the fly-back mechanism is confined to wheels L and M and their two levers.

When fitting the sweep second hand and the minute recording hand, the mechanism must be at zero. In this position the hands are fitted so that they point to zero.

When the push piece has been operated a third time and the ratchet wheel has been turned a third tooth, lever E is lifted by a pillar, which causes the opposite end of the lever to be moved away from wheel L and destroys the braking effect. At the same time, the two fly-back levers are released from the influence of the pillars, and under spring pressure they act on the two heart pieces, i.e., lever N on heart piece K, and the fourth lever acts on the minute recording heart piece. Almost instantaneously, both wheels spin around until they are stopped by the lever probes bearing in the lowest point of the heart-shaped cams. Both hands will now be pointing to zero and the chronograph cycle will be complete.

It has been mentioned elsewhere in the book that when dealing with complicated work it is essential that its function be studied before making any attempt to disassemble. Even when the mechanism is understood, it is good practice to spend a few minutes producing a rough sketch showing the relationship of the pieces. It may be good enough to rely on memory if the mechanism is going to be cleaned and assembled right away, but there is no guarantee that this will happen. A broken or damaged part may have to be replaced, or a piece may be dropped and lost. Under these circumstances the watch could remain in a disassembled condition for a few weeks, and by the time the work can be continued, the sequence of assembly may have been forgotten.

The layout of chronograph work varies considerably but in the majority of watches it is safer to remove the balance assembly and put it in a covered container away from harm. If the chronograph work can be removed with the movement in its case so much the better; otherwise the movement must be taken out.

Place the movement on a suitable rest and remove the hands. Turn the movement over and proceed to disassemble the chronograph work. Each lever should be placed in a separate compartment of a parts tray, along with its spring and screws. This is essential because the tension and length of the springs will vary, and quite often the screws may differ in length, although they may be cut with the same size thread. If a long screw is put into a threaded hole where a short screw belongs, the end of the screw may either touch the bottom and prevent the screw from being tightened, or harm another piece if the hole goes through.

The method of cleaning the pieces is the same as for any watch, but great care must be taken with wheels H, J and L. Their small size and their large number of very fine teeth make them delicate to handle. The teeth should be brushed carefully with a soft brush and then given a close inspection to ensure that every piece of dirt has been removed. The smallest piece of foreign matter lodged between two teeth could stop their function. If the teeth show signs of excessive wear, it is usually an indication that the wearer allows the chronograph mechanism to operate as a regular timekeeper. The wheels are not designed for continuous use, and the wearer should be advised accordingly.

The sequence of assembly usually begins with the ratchet wheel, followed by the wheel train, the levers and springs, then the hands, and finally the balance.

When assembly is complete, depress the push piece and make sure that the mechanism is functioning correctly throughout the cycle. Take note of the action of the finger piece on wheel L. Each time it gears with the minute recording wheel it should enter without harming anything else, engage the correct tooth, and pass through smoothly to the other side.

The position of the minute recording wheel in relation to the finger piece can usually be adjusted by the jumper spring, Fig. 96. If the spring is movable the wheel can be turned by altering its position. If the spring is fixed, polishing one of the faces of the jumper spring nose will turn the wheel. If face (a) is polished, the wheel will be turned clockwise, and if face (b) is polished the wheel will be moved counter-clockwise.

An adjustment is often provided for depth of engagement between wheels J and H. The pivot pin at G is sometimes made eccentric and can be turned, or the pivot is round but mounted eccentrically on a rotatable base. In either case, by moving the pin, the wheel J can be moved closer to, or away from, wheel H until the depthing is correct.

Oiling points are: ratchet wheel pivot, ratchet wheel pillar faces, pawl pivot, lever pivots, wheel pivots, bearing surfaces between springs and levers, and jumper ring noses.

XI

Self-Winding

THE EARLY SELF-WINDING OR AUTOMATIC WATCHES WERE NOT provided with keyless work, i.e., winding mechanism with stem and crown. Consequently, if the watch was not worn, the mainspring would run down and the movement would stop, and there would be no alternative method of rewinding.

Later models were provided with keyless work, so if the mainspring were allowed to unwind itself, rewinding it with a conventional stem and crown could quickly start the movement again.

The self-winding watch relies on the movement of the wearer's wrist to swing a weight which winds the mainspring. In some movements, winding takes place when the weight swings in one direction only. Other watches have a system that allows winding to take place in both directions of the swing. Many manufacturers have their own design but all are similar in operation. The energy in the swinging weight is transmitted through a ratchet and a train of gears to the ratchet wheel of the conventional winding mechanism.

Fig. 97 illustrates a system where the weight swings through an arc and winds in one direction only. Buffer springs are located at each end of the swing to absorb shock and so prevent damage.

Secured to the underside of the weight J is a pinion A which engages with a rack H. Located on the rack, but free to turn independently, is a ratchet wheel F and its pinion G. The pinion is

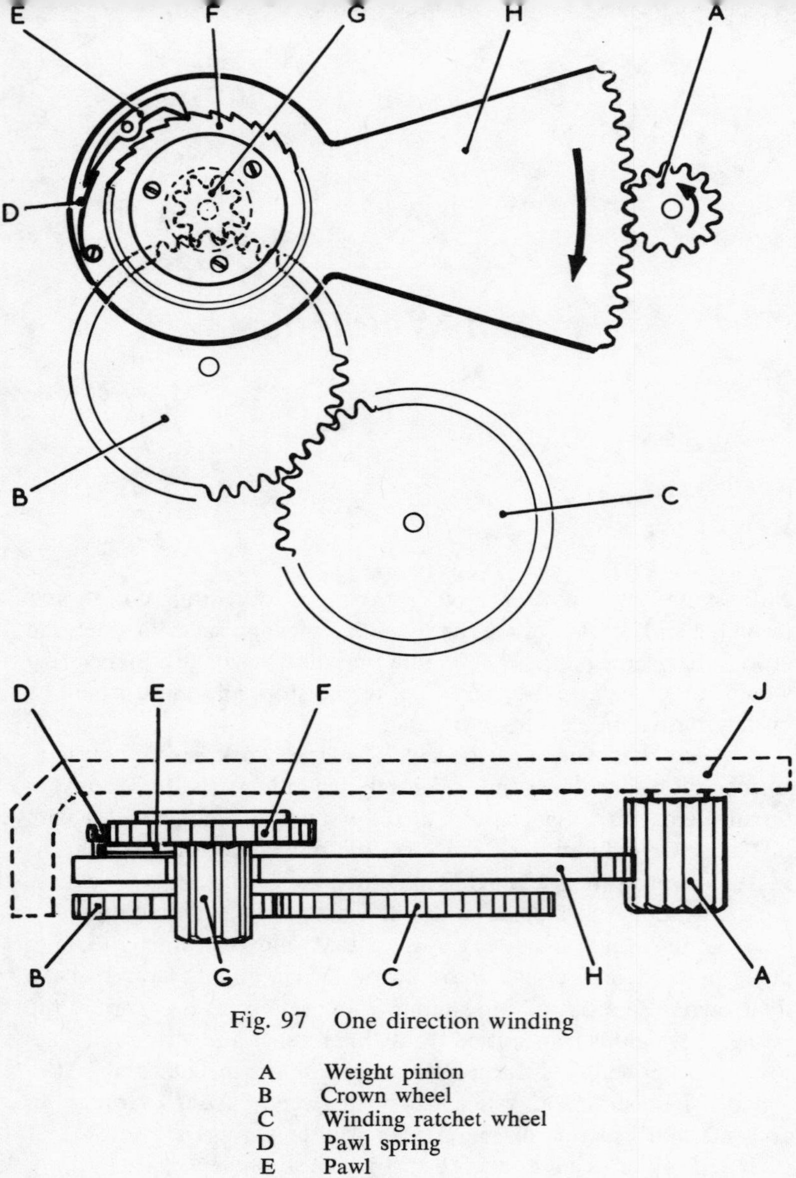

Fig. 97 One direction winding

A	Weight pinion
B	Crown wheel
C	Winding ratchet wheel
D	Pawl spring
E	Pawl
F	Ratchet
G	Ratchet pinion
H	Rack
J	Swinging weight

engaged with a crown wheel B which engages with a winding ratchet wheel C.

When the weight swings in a counter-clockwise direction, the weight pinion moves the rack in a clockwise direction. Mounted on the rack is a pawl E, which is held to the ratchet by a pawl spring D. As the rack moves around, the pawl turns the ratchet, and the ratchet pinion turns the crown wheel which is engaged with the winding ratchet wheel.

When the weight swings in the opposite direction, the pawl rides the ratchet until the weight returns, and the winding sequence is then repeated.

Later models were fitted with weights which rotated completely in both directions, thus doing away with the need for buffer springs. They still, however, wound in only one direction.

The next obvious step in design was to arrange these fully rotating weights, or rotors as they became known, to wind in both directions.

Fig. 98 shows a diagrammatic layout of a self-winding mechanism with a hand winding arrangement that is fitted in an Eterna-Matic watch, and Fig. 99 illustrates the principle of operation.

In Fig. 98 the rotor (1) is mounted on a ball-bearing (18) and held in place by a screw (17). The outer ring (16) of the ball-bearing is cut with teeth which engage the upper wheels (2) and (4) of the two pawl units.

Each pawl unit, Fig. 100, consists of two wheels mounted concentrically, one above the other, but which are free to rotate independently. Between the wheels is an intermediate disc that is machined to provide an inner ring and an outer ring of cam-shaped projections. Lying between these rings are two pawls that are bushed into the upper wheel. The two lower wheels are geared to each other.

The projections and the pawls are shaped so that when the upper wheel begins to turn and the pawls are carried around, they swing over and engage the projections. When this happens, the two wheels are locked together and rotate as one.

When the direction of the upper wheel is reversed, the projections cause the pawl to disengage and the upper wheel rotates independently of the lower wheel.

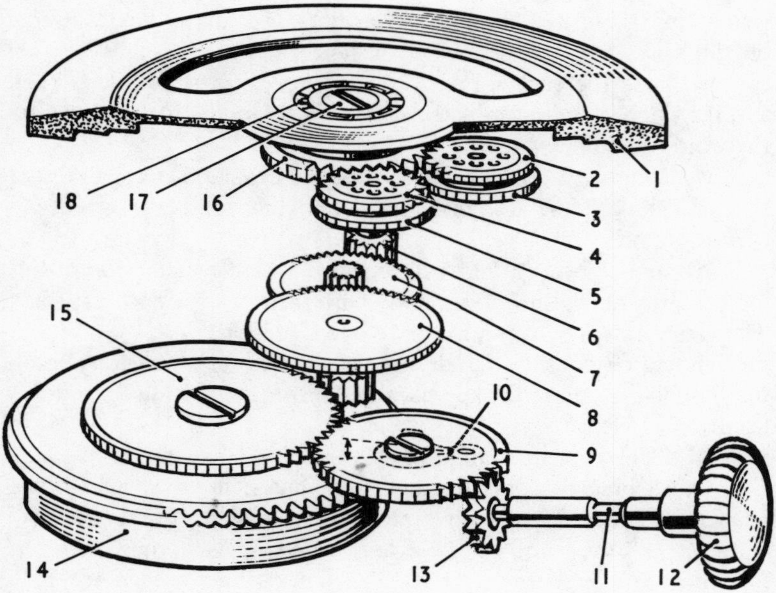

Fig. 98 Eterna-Matic diagrammatic layout

1	Rotor	8	Intermediate wheel
2	Upper wheel of auxiliary pawl unit	9	Crown wheel
3	Lower wheel of auxiliary pawl unit	10	Crown wheel yoke
4	Upper wheel of main pawl unit	11	Winding stem
5	Lower wheel of main pawl unit	12	Winding crown
6	Pinion of upper wheel 4	13	Winding pinion
7	Transmission wheel	14	Barrel
		15	Ratchet wheel
		16	Outer ring of ball-bearing
		17	Rotor screw
		18	Ball-bearing

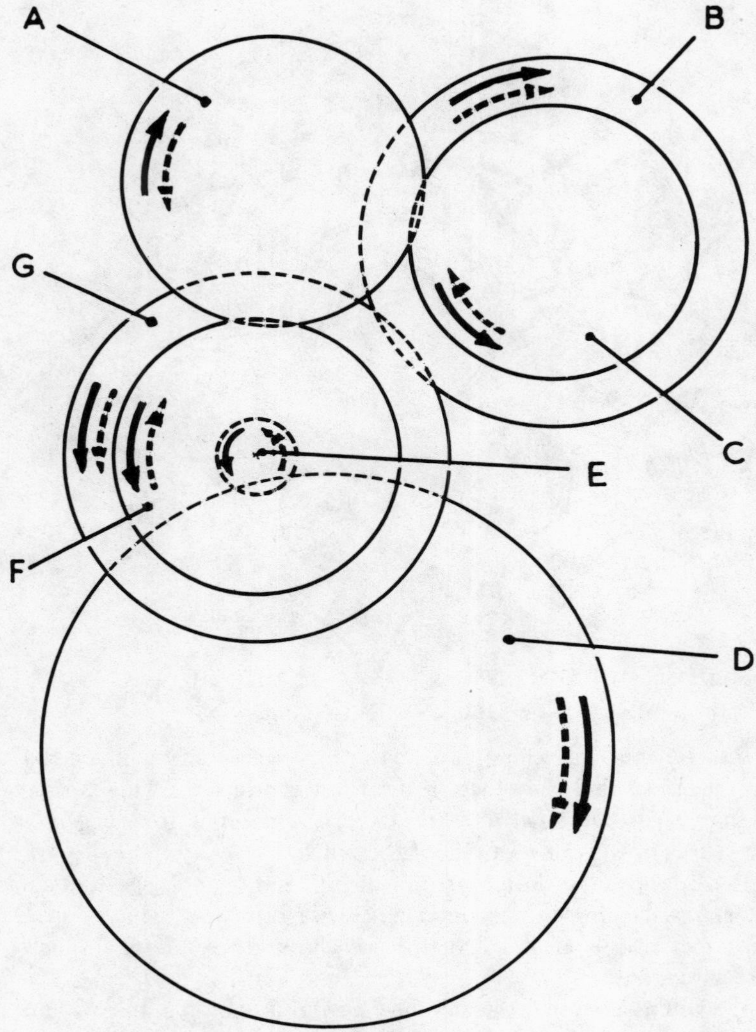

Fig. 99 Eterna-Matic principle of operation

A	Rotor wheel or toothed outer ring of ball-bearing
B	Lower wheel of auxiliary pawl unit
C	Upper wheel of auxiliary pawl unit
D	Transmission wheel
E	Pinion of main pawl unit
F	Upper wheel of main pawl unit
G	Lower wheel of main pawl unit

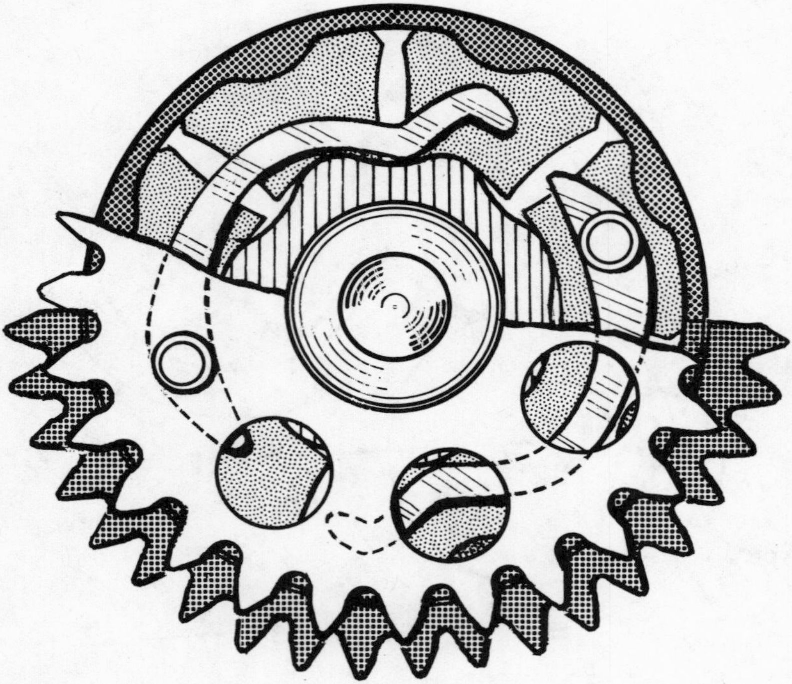

Fig. 100 Single pawl unit

Exactly the same thing happens in the other unit but the pawls are placed so that they face in the reverse direction. This is done so that when one pawl unit is locked, the other is free. Fig. 100 shows a cut-away view of one of the units.

If we follow the dotted arrows in Fig. 99, we will see that the rotor A is turning in a counter-clockwise direction, which causes the upper wheels C and F of the pawl units to rotate in a clockwise direction.

We will assume that the auxiliary pawl unit, i.e., the unit without a pinion, is locked and that wheels C and B are rotating together. If this is so, the main pawl unit must be free with wheels F and G unlocked.

Because the lower wheels B and G are geared to each other, the clockwise rotation of B causes wheel G to rotate counter-clockwise, and the pinion, which is secured to wheel G, must also rotate counter-clockwise. The transmission wheel D is geared to the pinion, and must therefore rotate in a clockwise direction.

When the rotor swings in the opposite direction, indicated in the drawing by an unbroken arrow, the upper wheels will change direction and rotate counter-clockwise. It is now the main pawl unit that is locked and wheels F and G rotate together.

Because wheel G is rotating counter-clockwise the pinion must do the same, and transmission wheel D must therefore rotate in a clockwise direction.

It follows then that regardless of the direction of rotor swing, the transmission wheel will always turn in the same direction.

The crown wheel is free to swing sideways within the limits of a yoke, so when the self-winding mechanism is an operation, the rotation of the ratchet wheel is able to push the crown wheel out of engagement.

Some Eterna-Matic watches have a slightly different arrangement, which is illustrated in Figs. 101 and 102. Instead of using two single pawl units as previously described, the mechanism is designed to function with one double pawl unit shown at Fig. 101.

Winding wheels G and F are free to rotate about winding arbor H, but pawl plate J is a tight fit and rotates with the arbor.

Secured to the underside of the rotor, Fig. 102, is wheel A which engages with reversing wheel B and upper winding wheel G. If the rotor is swinging clockwise, then wheels B and G will turn counter-clockwise.

Wheel B, which in effect is two wheels mounted one above the other and secured together, is also in mesh with lower winding wheel F, and turns it clockwise in the same direction as wheel A.

It will be seen from this arrangement that when the rotor swings, regardless of direction, winding wheels G and F will always rotate in directions opposite to each other.

Separating these two winding wheels is pawl plate J, which carries four pawls, two on top and two underneath, and which engage projections on the inner faces of the winding wheels.

With the rotor turning clockwise and wheel G turning counter-clockwise, the noses of the upper pawls engage the projections on wheel G, causing the pawl plate, with winding arbor and pinion, to be carried around with it. The lower pawls face in the opposite

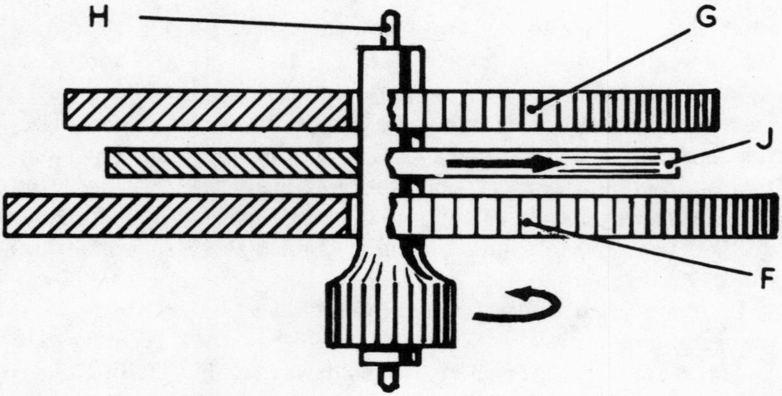

Fig. 101 Double pawl unit

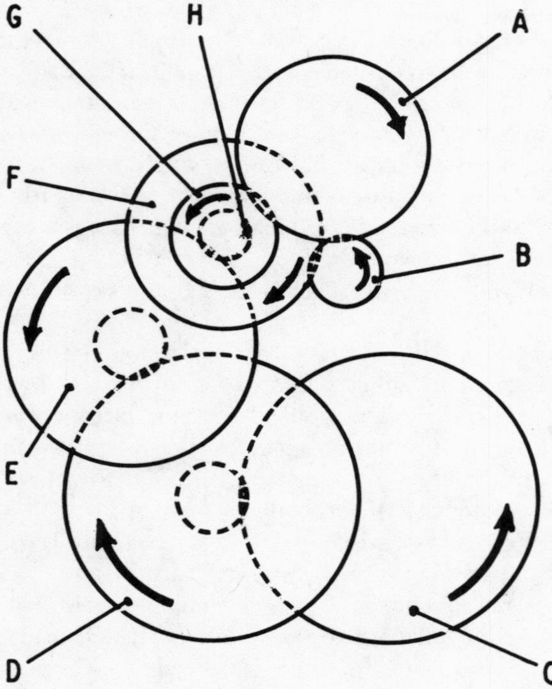

Fig. 102 Principle of operation with double pawl unit

A	Rotor wheel	F	Lower winding wheel
B	Reversing wheel	G	Upper winding wheel
C	Ratchet wheel	H	Winding arbor and pinion
D	Third winding wheel	J	Pawl plate
E	Second winding wheel		

direction and are moving backwards, and under these conditions are non-effective.

When the rotor changes direction it is winding wheel F that is turned in a counter-clockwise direction, and this time the lower pawls take up the drive while the upper pawls slip backward. Because wheel F has become the driving wheel, we find once again the winding arbor is being turned counter-clockwise.

In other words, the winding pinion turns counter-clockwise with every movement of the rotor, regardless of the direction of swing.

Winding wheels D and E are fitted to provide a reduction between winding pinion H and ratchet wheel C.

There are other methods of self-winding, some more simple in design than those described. One such method uses a pivotless wheel between the rotor wheel and the self-winding mechanism. The movement of the rotor causes the pivotless wheel to slide to one side and engage one wheel. When the rotor reverses direction, the pivotless wheel slides back again, and engages a different wheel. By this means, the rotor is allowed to fully rotate in either direction and engage the self-winding mechanism.

All present day self-winding watches employ a swinging weight or rotor to provide the motive power. A few minutes study while turning the rotor by hand will enable you to see how the design functions and what is the sequence of disassembly.

With some watches it is necessary to first remove the rotor by unscrewing a center fixing screw, or in some cases a locking plate or ring. The component parts are then removed separately from the movement.

The self-winding mechanism of an Eterna-Matic is removed as a complete unit by taking out either two or three screws, according to the model of the watch. The ball-bearing screw in the center of the rotor should not be disturbed.

This arrangement makes servicing the basic movement so much easier because there is no time spent in disassembling the self-winding mechanism.

After cleaning, the ball-bearing should be oiled with Eterna Rotoroil, which leaves a thin film of oil after evaporation. The pegs of the pawls in single pawl units need lubricating with watch oil but no oiling is required in double pawl units.

XII

Water-Resistant Cases

TO DESCRIBE A WATCH CASE AS BEING WATERPROOF IS MISLEADING. A more accurate description is water-resistant. The requirement of the majority of watches is to be resistant to the ingress of moisture from the air, and accidental but brief immersion in water. The practice of wearing a wrist watch when ones hands are in water is to be discouraged. Sooner or later water will find its way inside the case, usually entering at the winder, and rust will form on the movement, which frequently means parts must be renewed.

The three points of entry are the back, the crystal, and the winder. Watch manufacturers and case designers have introduced many different methods to ensure that their movements are kept free from moisture, including designs which have been patented. In principle, they are similar and all serve the same function.

There are watches made which are designed for prolonged use in water such as those worn by divers. These watches have to maintain accuracy while withstanding changes in temperature and pressure, in both fresh water and in the ocean. Such watches are specialized and can rightly be described as waterproof.

The majority of water-resistant cases are fitted with winding crowns that carry a plastic sealing ring in a recess under the head, Fig. 103. This ring fits close to the outside of the pendant and prevents moisture from passing inside the case.

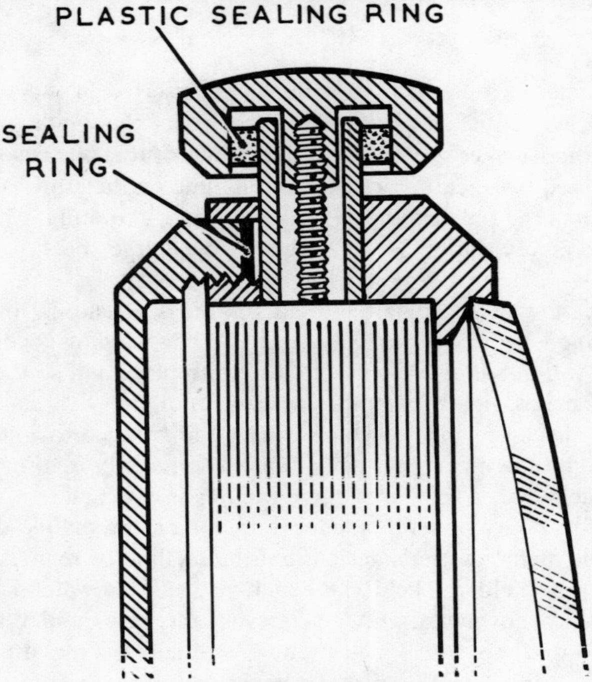

Fig. 103 Water-resistant crown

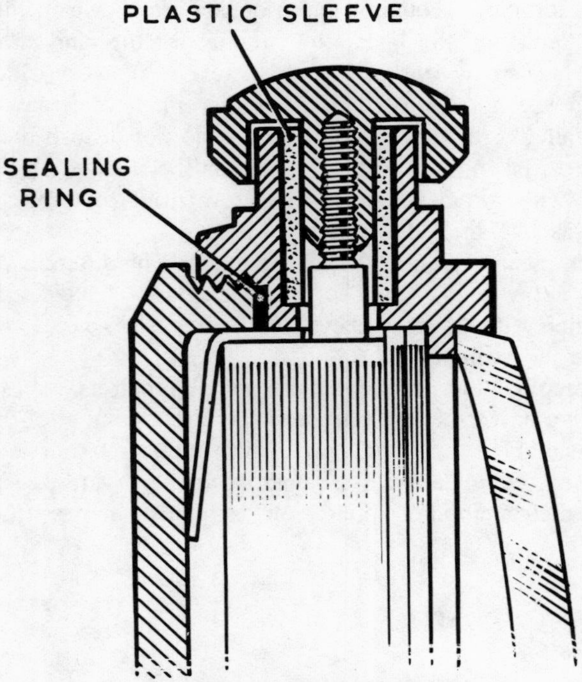

Fig. 104 Water-resistant crown

Other methods employ a plastic sleeve inserted in the pendant, Fig. 104, which fits snugly around the pipe of the winding crown. This method makes the crown rather stiff to turn, but usually this can be eased by smearing vaseline or lanoline on the crown pipe. In severe cases the pipe can be slightly reduced by rotating it between very fine emery paper. Never attempt to enlarge the hole in the sealing sleeve.

The nature of the material used for the seal causes it to pull on the pipe when the crown is operated. When winding is complete, this pull will not allow the click to drop into place and so the mainspring remains tightly wound. In this condition, the mainspring pulls on the barrel hook, which causes the balance to knock the banking and the movement gains time. The remedy is to move the crown backwards a few turns, after winding is complete.

Screwbacks can be threaded externally or internally, depending on the manufacturer's design, but the method of removal is the same. A case holder is held in a bench vise and the watch is placed on top, face downward. Four screws in the case holder are adjusted, so that they just touch the lugs that support the strap bars. A key is placed squarely over the back, and downward pressure is applied at the same time, which turns the key in a counter-clockwise direction. Some common types of key are shown in Fig. 105.

Before replacing the back, the original sealing ring should be discarded and a new seal fitted, but because of the very wide range of sizes in use it is possible that a replacement seal may not be readily available. In these circumstances, do not disturb the seal if it appears to be in good order, but smear the mating edge on the case back with soft beeswax. A mixture of four parts beeswax to one part vaseline, melted together, is ideal.

The comments made on the large number of different sizes of seal rings apply equally well to screwback keys. I have seen men at the bench with boxes of keys of different sizes none of which will fit the job at hand.

To overcome this problem, most watch repairers equip themselves with a universal case opener. There are a number of different models available; the more expensive ones are supplied in a wooden box, complete with a range of arms of various sizes. In principle, their method of application is similar. The basic tool consists of a strong metal ring into which are fitted screws or adjustable arms. The tool is placed over the case back, and the screws or arms are adjusted equally, until they have taken up a position of positive location in the grooves cut around the edge of the back or against the machined flats. The tool is now used as an ordinary key.

Fig. 105 Case back keys

The majority of watch cases are fitted with unbreakable plastic crystals. They are inexpensive, strong, and easy to replace. The edge of the crystal is slightly bevelled to fit into the groove of the bezel.

The correct method of fitting round crystals is to squeeze them between two metal blocks that are shaped concave and convex and which fit into each other. This arrangement is available in the form of a hand tool which is operated like a pair of pliers. The bezel is placed over one arm, the plastic crystal is inserted between the blocks and the tool closed firmly over it. By squeezing the plastic crystal the diameter is made smaller. This allows the bezel to be positioned around the crystal, and as the pressure on the tool is reduced the crystal tries to resume its former diameter but is held under slight tension in the bezel groove.

Another method of inserting and removing round crystals is to use a crystal inserter and remover tool, sometimes referred to as a crystal lift. The tool is cylindrically shaped and consists of a number of steel prongs formed in a circle, the diameter of which is adjustable to fit and grip the outside of the crystal. The advantage of this tool is that it gives the ability to remove and replace a crystal without removing the movement from the watch case.

It is wrong to try and force the crystal into the bezel by hand.

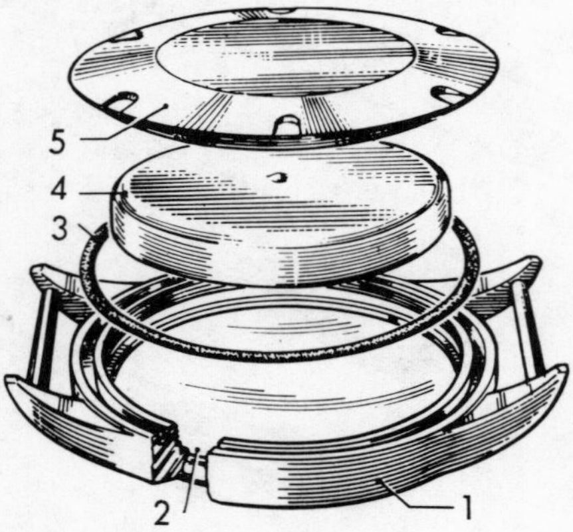

Fig. 106 Screwback case

1 Case front
2 Crystal (armoured)
3 Case gasket
4 Movement cover
5 Case back

This will remove the sharpness of the bevelled edge and result in a loose fit.

A more sophisticated tool is a small press that screws to the bench but the principle of operation is the same.

Fig. 106 shows a case with a screwback and a press-in type of movement cover. When the back has been removed, the watch is held so that both thumbs are on the crystal. By pressing hard, the movement and its cover will be pushed from the bezel, along with the sealing ring and crystal. This operation must be carried out close to the bench top to prevent the movement falling to the floor.

Some manufacturers fit snap-on crowns and stems to their water-resistant watches, Fig. 107. Before the movement can be removed

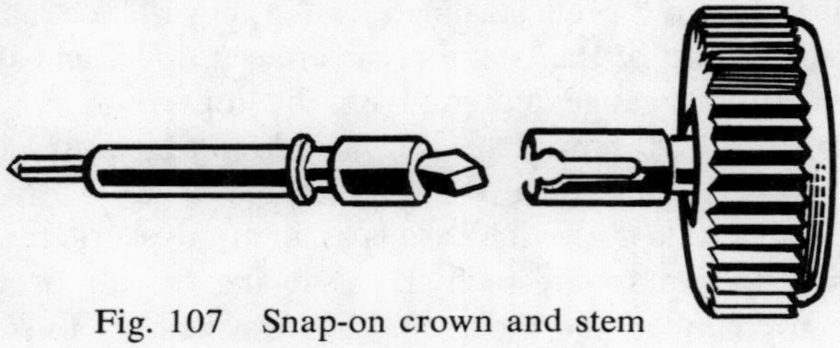

Fig. 107 Snap-on crown and stem

the crown must first be separated from the stem. This is done by wrapping a small piece of chamois leather or cleaning cloth over and under the crown for protection, and then gently closing the jaws of a pair of top cutting nippers beneath the crown. The lever action will ease the crown from the stem. To replace the crown the slot must first be engaged with the tongue on the end of the stem, and then a push or light tap will snap the two together.

There are two tests to which water-resistant cases can be subjected. One is known as the pressure test and the other the vacuum test.

In the pressure test the watch case is assembled complete with winding crown but without the movement. The case is suspended in a cylindrical glass pressure chamber which is mounted on a metal base. The cylinder is filled with water and a metal cap is screwed down on the top. A plunger in the base is screwed inward which increases the internal pressure. This pressure is indicated on a gauge which is mounted on the side of the base. The usual pressure at which these tests are conducted is 4.3 lbs per square inch, which is equivalent to a depth of ten feet.

The case is left in the chamber for one hour. After this time, the plunger on the test equipment is screwed back to relieve the pressure, the top is removed and the watch case lifted out and dried. The case is opened and inspected for signs of moisture. If the case shows no sign of water leakage the movement can be fitted but having disturbed the case there is no guarantee that it will be resistant to moisture after re-assembly.

For more practical purposes the vacuum test is considered to be the better. The case is assembled complete with movement, and suspended in a cylindrical glass vacuum chamber. Water is put into the chamber to a predetermined level to ensure adequate air space above. The top of the chamber is screwed down and the air withdrawn until a gauge mounted on the top cover registers 8.776 inches or 222.90 mm of mercury, both of which are equivalent to 4.30 lbs per square inch or a depth of ten feet.

The air inside the watch case is at an atmospheric pressure equal to 14.7 lbs per square inch. By reducing the air pressure in the vacuum chamber, the air inside the case will try to force its way out and in so doing will produce air bubbles.

Observe the case closely and if bubbles appear, note from which position they emerge. Remove the watch immediately and open the case to make sure that water did not enter. Reseal the case and test again. When there are no bubbles the case can be considered water-resistant for everyday purposes.

DEPTH OF WATER IN FEET	LBS. PER SQ. INCH PRESSURE	INCHES OF MERCURY	MILLIMETERS OF MERCURY
1	0.43	0.878	22.29
2	0.86	1.755	44.58
3	1.29	2.633	66.87
4	1.72	3.510	89.16
5	2.15	4.388	111.45
6	2.58	5.266	133.74
7	3.01	6.143	156.03
8	3.44	7.021	178.32
9	3.87	7.898	200.61
10	4.30	8.776	222.90
11	4.73	9.654	245.19
12	5.16	10.531	267.48
13	5.59	11.409	289.77
14	6.02	12.286	312.06
15	6.45	13.164	334.35
16	6.88	14.042	356.64
17	7.31	14.919	378.93
18	7.74	15.797	401.22
19	8.17	16.674	423.51
20	8.60	17.552	445.80
21	9.03	18.430	468.09
22	9.46	19.307	470.38
23	9.89	20.185	512.67
24	10.32	21.062	534.96
25	10.75	21.940	557.25
26	11.18	22.818	579.54
27	11.61	23.695	601.83
28	12.04	24.253	624.12
29	12.47	25.450	646.41
30	12.90	26.328	668.70

XIII

Electronic Watches

THE EARLIEST WATCHES WORKED BY USING A MAINSPRING WHICH gave power to a train of gears. The mainspring was allowed to run down under the control of an escapement and oscillating balance. The oscillations were regular and were counted by the gears, which in turn operated the hands. The first of these movements were supplied with separate keys which were inserted into the back of the movement for winding the mainspring.

Next came the keyless watch. The keyless work, as it is known, introduced the winding stem and crown with which we are familiar.

A further development was to retain the keyless work but introduce a swinging weight or rotor which wound the mainspring with each movement of the wearer's wrist. This was known as the automatic or self-winding watch.

Throughout these changes it was only the method of winding the mainspring that altered the design of the movement remained basically the same.

Then came the introduction of an electrically operated movement powered by a small battery. This was the beginning of a complete breakaway from what had been traditional design, and which many people thought could not be improved.

The first battery operated movements were fitted with a set of make-and-break contacts which were made to open and close mechanically by the oscillations of the balance. When the contacts closed, current was allowed to flow through a coil setting up a magnetic field around itself. This magnetic field repelled a permanent magnet that was part of the balance assembly, and caused the balance to oscillate in the reverse direction.

When this happened, the circuit breaker opened and the magnetic field was destroyed, but the action of the balance moving in the opposite direction closed the breaker again and the cycle was repeated. It was these alternate impulses that maintained oscillation of the balance.

The oscillating motion of the balance was transferred into rotary motion by an index finger that acted upon an index wheel, turning it one tooth at a time. The index wheel was the first of a train and consequently all the gears were set in motion. Sometimes a pawl finger was used to hold the index wheel steady between impulses, and prevent it moving in reverse.

The next major improvement was the introduction of an electronic circuit. Instead of using a balance to operate the index finger, this function was now performed by a tuning fork vibrator, and circuit breakers for switching battery current were discontinued in favor of transistors.

Figs. 108 to 111 illustrate these progressive stages in diagrammatic form.

The next improvement, and without doubt the most dramatic, was the use of a resonator of quartz crystal in place of the tuning fork vibrator.

Quartz is a rock-forming mineral composed of silica which is found in hexagonal crystals which are clear and colorless when pure. When quartz is subjected to an electrical impulse it vibrates with remarkable stability at precisely 8192 times per second, every second. With such high frequency vibrations, extreme accuracy is possible and these movements maintain correct time to within five seconds per month.

But with all these changes, the gear train has remained. And now in this age of space travel, technological and scientific progress has enabled designers to produce the most futuristic watch of all. It is an electronic watch that retains the quartz crystal resonator but dispenses with hands, dial and gear train in favor of digital indication by using luminous numerals. With such an arrangement there are no moving parts.

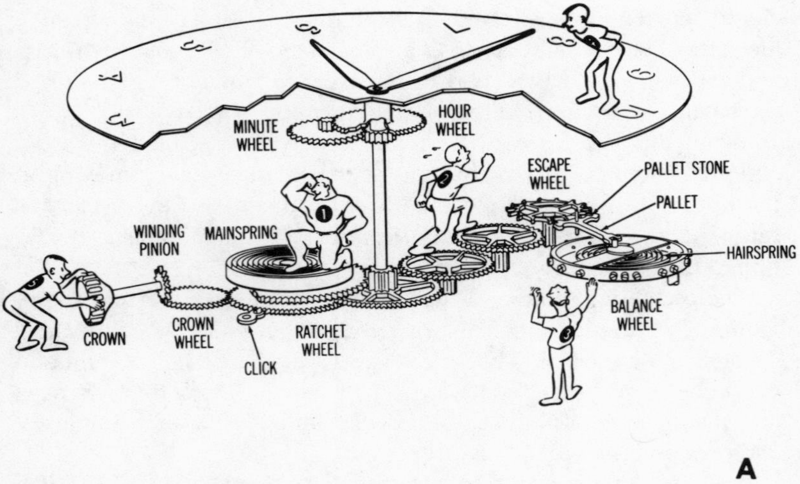

Fig. 108

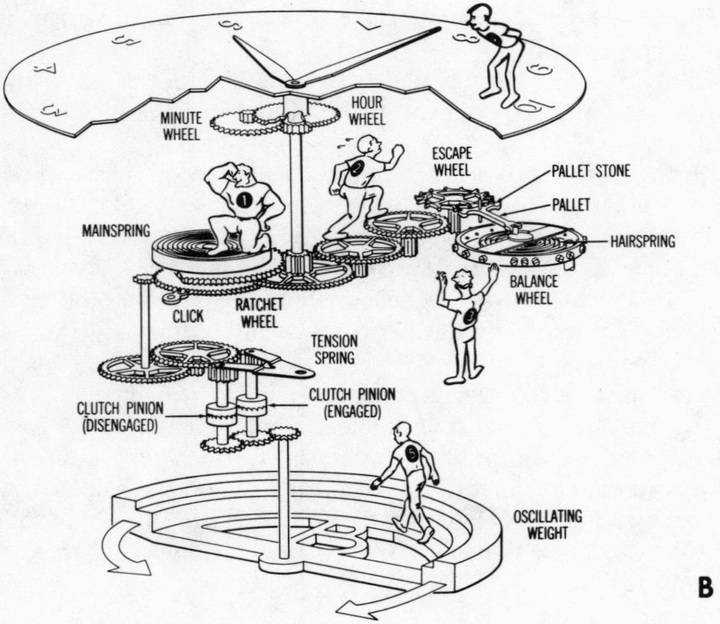

Fig. 109

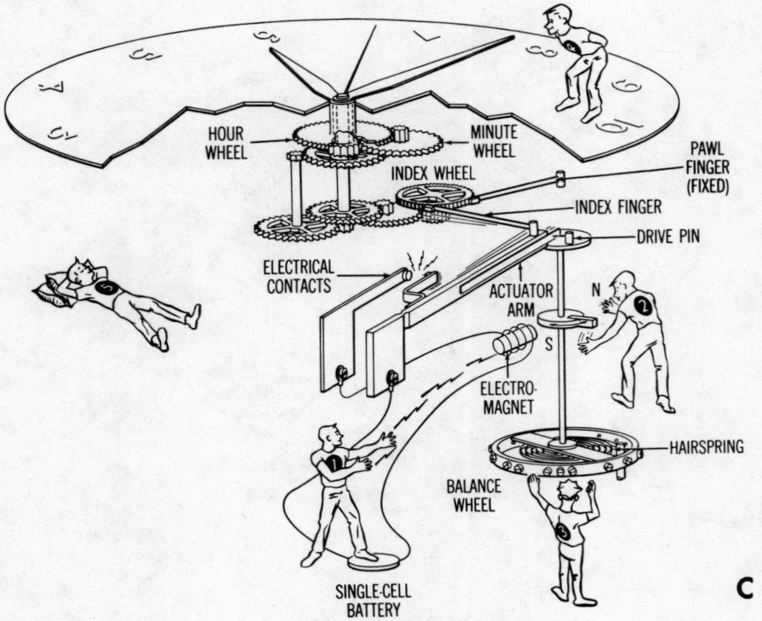

Fig. 110

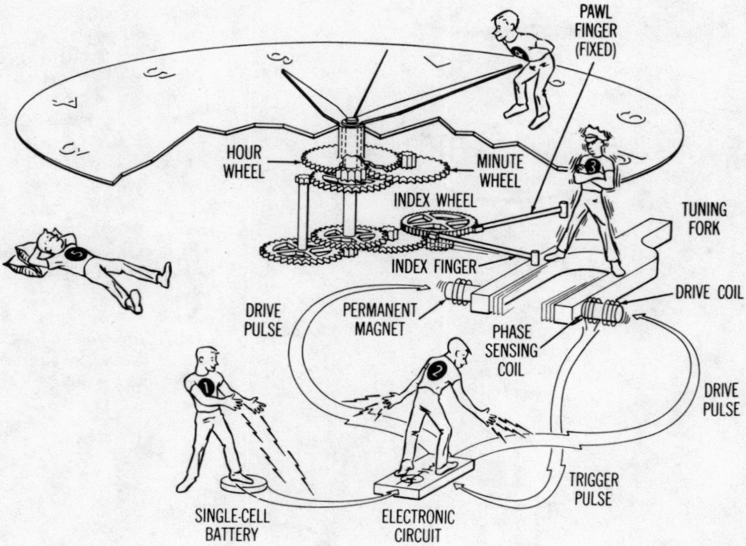

Fig. 111

the TIMEX model 69 movement (exploded view)

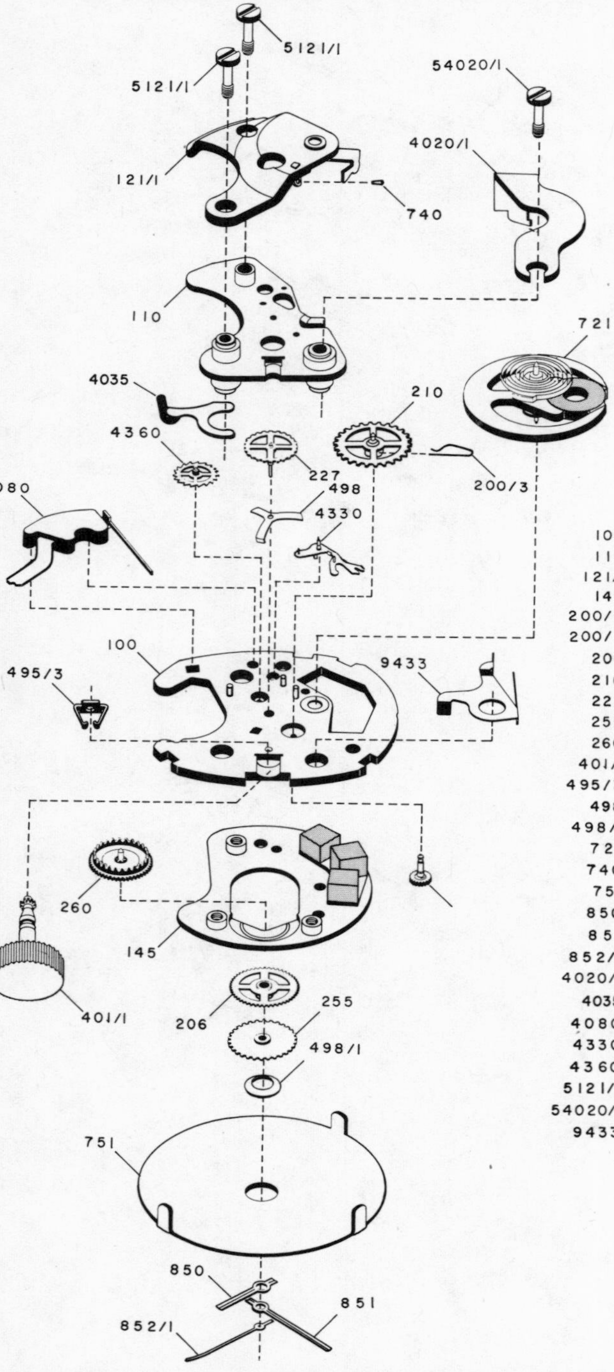

100	PLATE ASS'Y.
110	TRAIN WHEEL BRIDGE ASS'Y.
121/1	BALANCE BRIDGE ASS'Y.
145	DIAL REST
200/2	FRICTION PINION
200/3	FRICTION SPRING
206	CENTER WHEEL ASS'Y.
210	THIRD WHEEL ASS'Y.
227	SECONDS WHEEL ASS'Y.
255	HOUR WHEEL ASS'Y.
260	MINUTE WHEEL ASS'Y.
401/1	SETTING STEM WITH CROWN
495/3	STEM BRACKET
498	FRICTION WASHER
498/1	HOUR WHEEL WASHER
721	BALANCE ASS'Y.
740	HAIRSPRING WEDGE PIN
751	DIAL
850	HOUR HAND
851	MINUTE HAND
852/1	SWEEP SECOND HAND
4020/1	SHUNT BRIDGE
4035	ENERGY CELL SPRING
4080	CONTACT SPRING ASS'Y.
4330	INDEX LEVER ASS'Y.
4360	INDEX WHEEL ASS'Y.
5121/1	BALANCE BRIDGE SCREW
54020/1	SHUNT BRIDGE SCREW
9433	SET SPRING

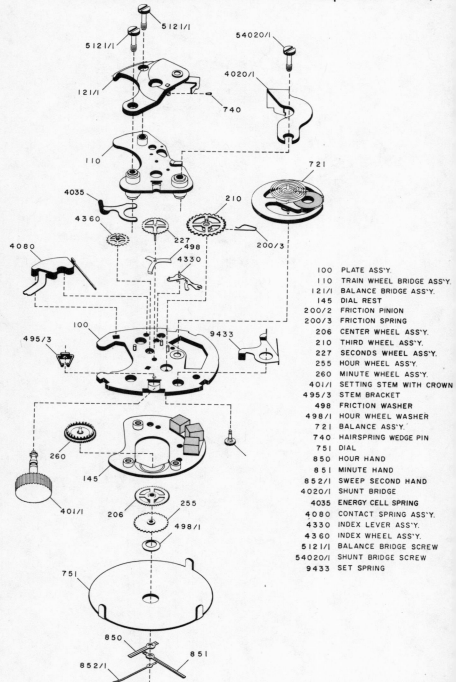

Fig. 112 The Timex model 69 movement (exploded view)

The Electric Watch

Fig. 112 shows an exploded view of the movement in a Timex ladies' watch. The action is based on the principle of using an electric circuit make-and-break to cause magnetic impulses with which to oscillate the balance.

The battery is housed in the back of the watch and can be replaced without opening the case. Pry off the battery cover by using a knife blade. Turn the watch over and the battery will drop out. Make sure the replacement battery is fitted the correct way around. It is the negative (−) end of the battery that is inserted into the watch. The electric circuit is completed only when the cover has been snapped back into position.

The hands are set in the usual way by pulling out the stem and rotating the crown. The action of withdrawing the stem stops the oscillation of the balance by mechanical means, and at the same time the electric circuit is broken. When the stem and crown are returned the mechanical brake is removed, the electric circuit is closed and current from the battery sets the balance in motion again.

The movement can be disassembled and overhauled without the need for special tools or a knowledge of electricity.

After the battery has been taken out, remove the crystal and reflector ring using a conventional crystal lift.

Next, remove the stem and crown. The stem is held in place by a stem bracket, and mounted on the stem is a sliding sleeve. Pull the stem into the set position and insert a screwdriver blade adjacent to the 3 o'clock position on the dial. Hold the movement in place and push the sleeve along the stem toward the center. This will spread the bracket and allow the stem to be withdrawn.

The sleeve is loose on the stem and it can slide into the pendant. If this happens, move the slide along the stem with a pair of tweezers so that a screwdriver blade can be inserted behind the slide.

After the stem and crown have been withdrawn the movement can be removed from the case. At this stage it may be necessary to check the operation of the movement. To do this, the stem and crown must be refitted. Hold the movement steady in the case, push the stem into position and, at the same time, turn the crown so that the stem pinion will engage with the minute wheel.

The battery must now be held in position in the movement. This is done by using a retaining spring which can be obtained from any Timex material supplier.

Make sure that both ends of the battery are making electrical contact with the movement and if the battery is suspect, fit a new one.

To disassemble the movement first remove the battery. Then rotate the balance to lock position to avoid damaging the coil when removing the shunt.

Take out the shunt bridge screw and carefully raise the shunt bridge at the screw end. The opposite end is held by a tab that is hooked over the plate. If the shunt bridge is now moved in the direction of the tab, it will be freed from the plate and can be lifted clear. The shunt bridge screw should be replaced and tightened to prevent the train wheel bridge being disturbed when removing the balance bridge.

The action of the balance and the contact spring can now be studied.

By turning the balance slowly from its position of rest, Fig. 113, it will be seen that at a predetermined position the contact pin touches the contact spring. Immediately after this happens, current from the battery flows through the balance coil setting up a magnetic field. This field is in opposition to the magnetic field surrounding the permanent magnet and so, following one of the laws of magnetism that states that unlike poles attract and like poles repel, an impulse is given to the coil which pushes the balance in reverse.

After the midway position has been passed, Fig. 114, the cycle starts up again and the balance will continue to oscillate under the influence of the impulses as long as current is available from the battery.

The contact spring is very delicate and must be handled with the same care given to a balance spring. The length is fixed during manufacture and must not be altered. When at rest, the spring should be pointing exactly to the center of the balance staff, and at the same time be horizontally positioned centrally between the impulse disc and the balance wheel.

Carry out any necessary positional adjustments with a pair of tweezers, but do so with care. Avoid excessive bending or any other deformations of the spring. The contact end should be left untouched. It must not be scratched, polished or handled in any way.

It is essential that contact be made between the contact pin and the contact spring at a precise position of the balance. To achieve this, a contact jewel, whose function is to limit the physical contact between the pin and the spring, is fitted.

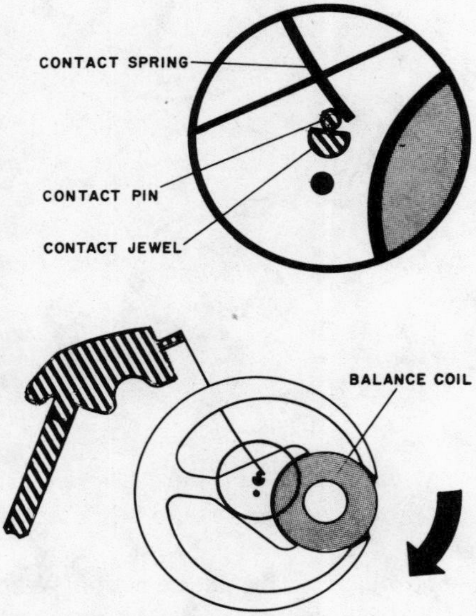

Fig. 113

Oil is a bad conductor of electricity and therefore the contact spring, contact pin and contact jewel must be quite dry.

Unpin the balance spring from the regulator. Remove the balance endshake screw and the balance bridge screw and lift off the balance bridge assembly. The balance is now exposed and may be lifted out. Take care not to damage the coil; it is wound with extremely fine copper wire and will not withstand handling with tweezers.

Hold the train wheel bridge and plate together, and remove the shunt bridge screw, the dial rest, and the minute wheel assembly. Holding these two parts together will prevent the gear train parts from being attracted to the magnet when the train wheel bridge is

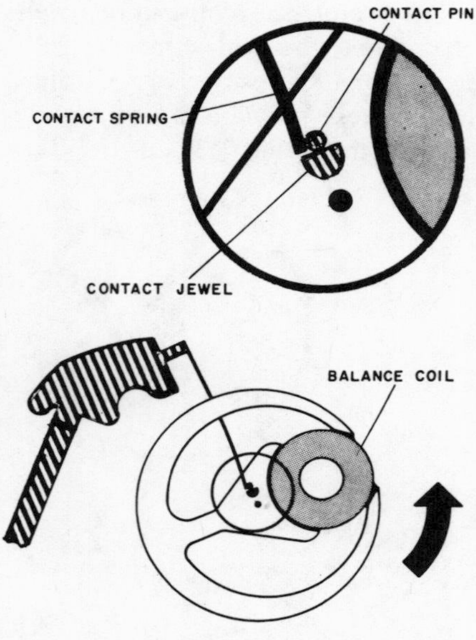

Fig. 114

removed. Place the movement on a stand and carefully remove the train wheel bridge.

Figs. 115 and 116 illustrate the index action. When the balance moves clockwise the impulse pin engages with the fork of the lever and moves the lever counter-clockwise.

Pin D engages with the index wheel, moving it forward approximately three quarters of one tooth. Magnets A, B and C attract the tips of the index wheel teeth and move the wheel forward the remaining quarter tooth. Magnet C also attracts the draw lever on the index lever and holds the index lever in position and magnet B also acts as a banking pin for the index lever.

In the event that the index wheel is malpositioned, pin E will turn it to its correct position so that the next engagement of pin D will advance the index wheel correctly.

A friction spring connects the third wheel and the friction pinion, and it provides the friction necessary for setting the hands. The two can be separated by lifting the pinion with tweezers and removing it from the spring.

To reassemble the third wheel, place the friction pinion on a flat plate and lower the movement plate over the pinion. The third wheel, complete with friction spring, is then snapped onto the post of the pinion.

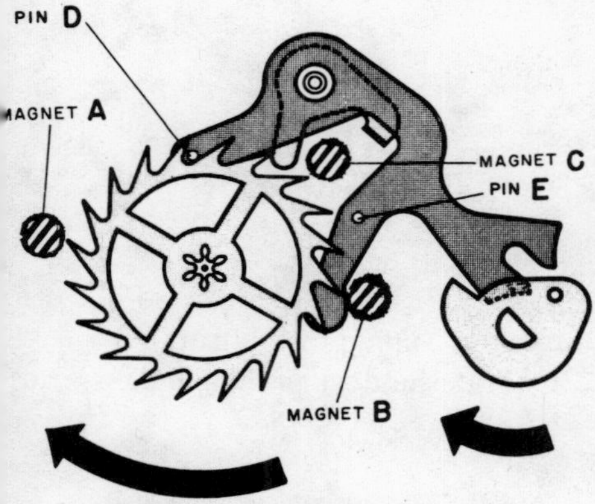

Fig. 115

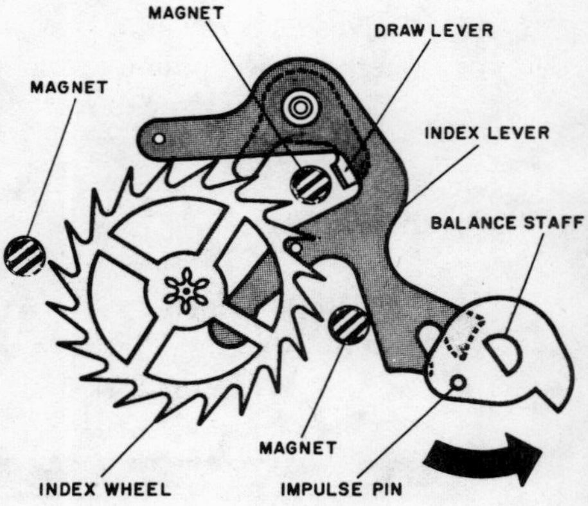

Fig. 116

Between the second wheel and the plate is a domed friction washer, which is prevented from rotating by a bent tab located in a slot in the plate. If the washer has become flattened, it will cause erratic movement of the second hand. Too much curvature to the washer will cause drag to the movement.

When the balance bridge and balance have been removed, there is no need for further disassembling if the movement is to be cleaned and oiled only. Cleaning can be carried out by hand or by machine but to prevent damage to the balance coil it is better that the balance be cleaned separately.

A note of caution: The balance bridge screw close to the 9 o'clock position also acts as a balance endshake adjuster. When the screw is tightened the endshake is reduced, and slackening the screw increases the endshake. Care must be exercised when tightening the screw to prevent breaking the balance staff pivots.

Close inspection frequently discloses metal particles adhering to the magnets; these particles can be removed with adhesive tape.

After cleaning, oil should be applied to:

balance bearings
impulse pin
index lever pivots
index wheel teeth
all wheel pivots
center wheel pinion to friction washer
minute wheel to dial rest

Use a very light, silicone-free grease on the stem where it is held by the bracket.

Tuning Fork Watch

Most of us are familiar with the musicians' tuning fork illustrated in Fig. 117. When the fork is struck it vibrates and when the handle

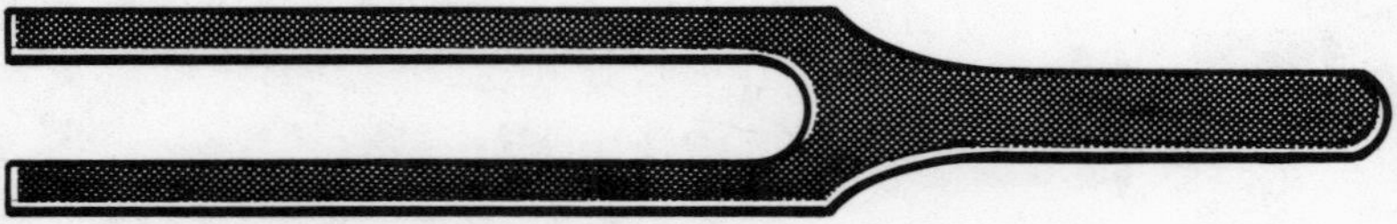

Fig. 117 Tuning fork

is held against something hard and solid, the vibrations produce an audible note. The frequency of vibrations depends on the dimensions of the fork, and so different sizes of fork produce different notes.

The principal characteristic of a tuning fork is its ability to produce very high frequency vibrations. It is because these vibrations are so numerous that great accuracy in timekeeping is possible.

To transform the vibratory motion of the fork into useful rotary motion, an index finger is coupled to one of the arms or tines of the fork. The other end of the index finger engages the teeth of an index wheel. With each complete cycle of a vibration, the index finger moves the index wheel one tooth. The pinion of the index wheel is geared to the wheel train so the wheel train is set in motion.

The musicians' tuning fork vibrates by mechanical means, i.e. striking it by hand, but the tuning fork in an electronic watch vibrates by electro-magnetic forces from energy supplied by a small 1.3 volt mercury battery (power cell). If the watch is held to the ear the hum of the fork can be heard.

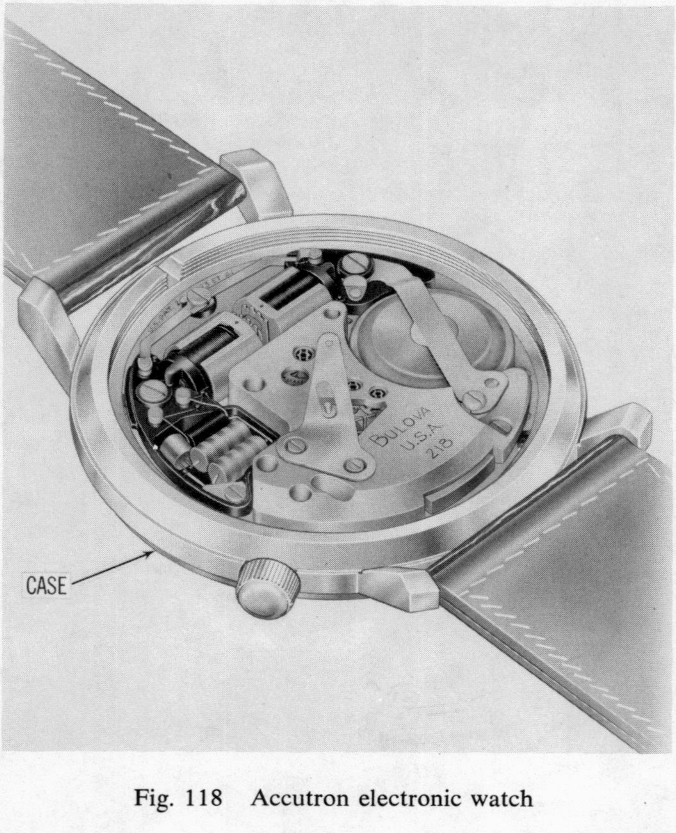

Fig. 118 Accutron electronic watch

Fig. 118 illustrates a Bulova Accutron tuning fork watch with the case back removed. The upper part of the fork can be seen in the 12 o'clock position. The complete tuning fork is shown in Fig. 119. It is made from a self-compensating alloy similar to that

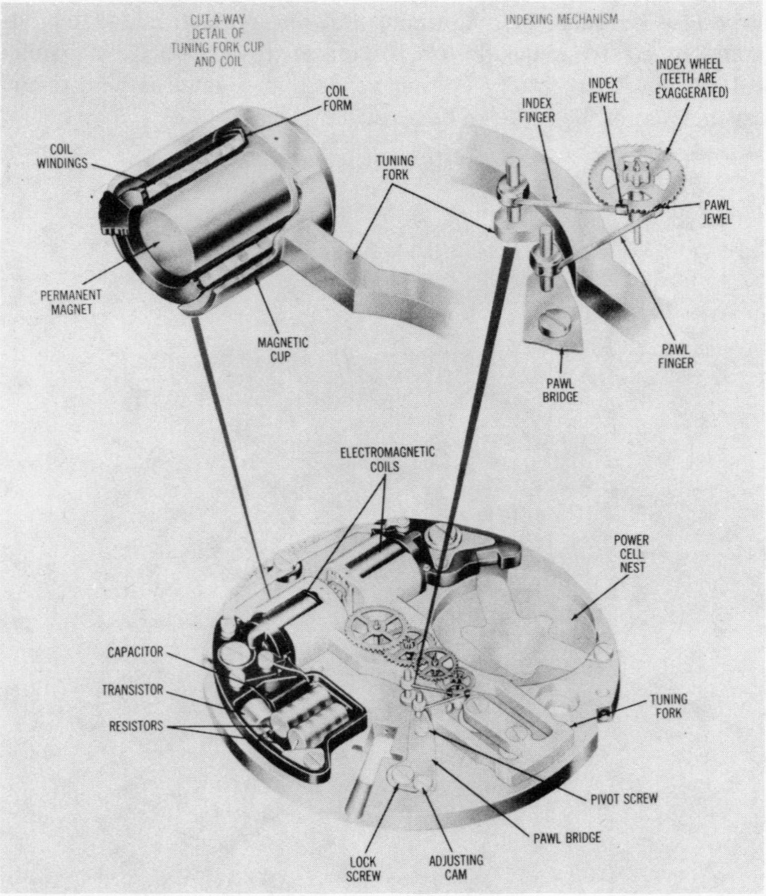

Fig. 119 Accutron tuning fork and index

used for hairsprings, and the rate is therefore unaffected by normal changes in temperature. In Fig. 119 the tuning fork can be seen in position on the pillar plate. The long arms of the fork are known as tines, and at the end of each is a cylindrical permanent magnet surrounded by an iron magnetic cup. A strong magnetic field exists in the space between the cup and the magnet.

The transistorized electronic circuit releases small pulses of electrical energy from the battery, and passes them to the tuning fork drive coils that are supported by two plastic moldings mounted on the pillar plate. These coils are tubular and extend into the magnetic cups, and therefore cut the lines of magnetic force set up by the permanent magnets. The coils are concentrically positioned with clearance all around, so that the fork is free to vibrate without touching the coil windings.

One of the coils is wound in two parts with approximately three times as many windings in one part as in the other. The larger of the two parts is used as a drive coil to operate the fork, and the remaining windings, known as the phase sensing coil, determine when a pulse of electric energy is passed to the drive coils.

When an electric current passes through the windings, the coils become electro-magnets and, depending on polarities, will either attract or repel the magnetic cups. The fork moves, and in so doing the coils are made to cut the lines of magnetic force, and an alternating voltage is induced in the coils. The direction of the fork is reversed, the cycle repeats itself, and the fork continues to vibrate. A circuit diagram of an Accutron watch is shown in Fig. 120.

To service or overhaul an electronic watch does not call for a

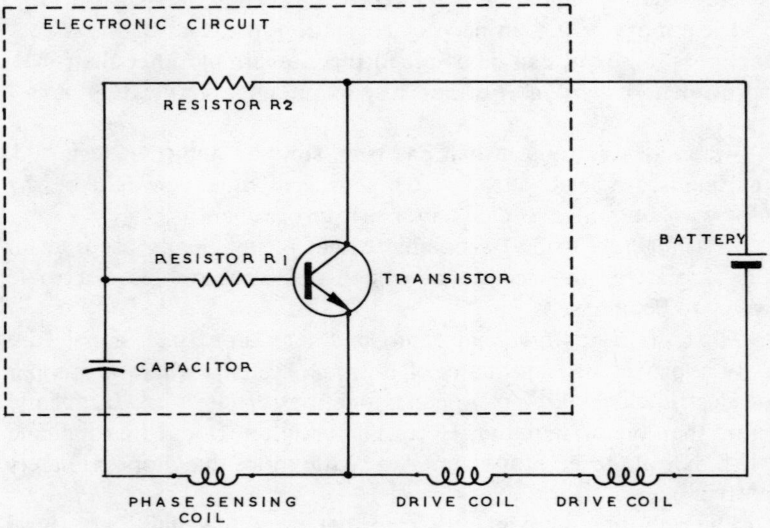

Fig. 120 Circuit diagram

knowledge of electronics on the part of the watch repairer. If a fault develops with an electronic component or in the electronic circuit, it usually means a unit or component replacement. With the correct test equipment the checks are simple and straightforward.

The indexing mechanism is the most delicate part of the movement and must be handled with extreme care. The index wheel fitted to the series 214 Accutron is so small that it is impossible to carry out a functional examination of the mechanism without the aid of a microscope. The diameter of the wheel is only 2.40 mm and yet 300 teeth have been cut in the circumference. The microscope needs to have a magnification of 20 to 30 diameters, and at least a 2 inch working distance. The necessity of a microscope is not apparent with the later series 218, illustrated in Fig. 118, and with other makes of watches. Many watch repairers are able to work with a loupe.

The remaining parts of an electronic watch conform to conventional design and require the same attention as any other movement.

In ordinary watches, quite large gains or losses can be corrected by regulation alone. This is not possible with electronic watches. The amount of regulation is limited to a very few seconds per day, and if a movement is in need of correction in excess of the manufacturers' limits, it can be assumed that something other than maladjustment is wrong. Further inspection and rectification is required.

When an electronic watch has been stopped and then restarted, the hands are set to any reliable source of time as would be the case with any watch or clock after service or repair.

It is not easy to find a timepiece that is sufficiently accurate to the second for the occasional owner who requires his electronic watch to be precise.

Most of the commercial radio and television time signals are only approximate, and many of the time systems where a number of electric clocks are governed by one master clock are frequently more than one minute out. Even the pendulum clocks that indicate "U.S. Naval Observatory Time" are little more than approximately correct.

The only precise signals that are commonly available are those given by the telephone service and those which are transmitted by short wave radio stations operated by government departments.

In addition to the usual watch repairers' tools, there are a few special items that are needed in order to work on electronic watches. These tools are available directly from the manufacturers, and include an electric test set, a movement holder, a case back wrench, and an index mechanism adjusting tool. A simplified fault finding chart is shown at the end of the chapter.

Renewing a Battery—Watch case design varies considerably between manufacturers and models, and the procedure for renewing batteries varies accordingly. To change a battery in an Eterna Sonic water-resistant watch 11T-1550 it is necessary to remove the movement from its case. The hand set is a two piece stem and has to be separated. The outer half, complete with crown, is lifted off by using a pair of hand removing levers, and the inner half of the stem is then pushed back down the tube.

Remove the four case screws which are located in the bracelet horns of the case. Lay the watch on the workbench, dial face up, and with finger pressure of the left hand press down lightly on the crystal. At the same time very carefully remove the bezel with the right hand, taking care not to disturb the crystal. Turn the watch over and keep it pressed down by holding the edge of the plate. Carefully lift the back of the case from the movement.

While the crystal remains in position, there can be no possible contact with the sweep second hand and consequently the index mechanism will be protected from damage.

Some watch cases have a small threaded cover in the case back that can be unscrewed with a small coin. By removing the cover and then turning the watch over the battery will drop out.

If the watch has a screw back case that must be removed to expose the battery the correct wrench should be used.

Having removed the battery, the electrical contacts should be inspected for cleanliness to ensure that perfect electrical contact will be made when the new battery is fitted.

The voltage of the new battery should be checked before fitting, and, providing the meter reading indicates a voltage of between 1.25 and 1.45, the battery is fit to use. The test instrument should be either a high resistance voltmeter with not less than 10,000 ohms per volt sensitivity, or the test set that is provided by the manufacturer.

When testing, make sure that the instrument needle remains steady. A wavering needle is an indication that there is a poor electrical connection which produces an inaccurate reading.

It is very important that the type of battery recommended by the manufacturer for the particular model of the watch concerned, be used. Batteries are not necessarily interchangeable even though their dimensions may be identical.

Never use a hearing aid battery because there is a danger that the electrolyte might leak from the battery case when the battery nears the end of its life. A leakage would cause serious damage to the movement from acid corrosion.

If the positive (+) and the negative (−) ends of a battery were to be held between a pair of metal tweezers there would be an electrical discharge which would reduce the useful life of the battery. Always use tweezers of a non-conductive material such as nylon, or metal tweezers with insulated ends, or ones fingers.

After the new battery has been fitted, a light tap to the watch case will start the tuning fork vibrating.

Regulation—An electronic watch can be regulated to maintain accurate time, providing the gain or loss is not in excess of the tolerance within which the regulators are designed to operate.

With the back of the case removed, regulation is carried out on the tuning fork by rotating a pair of weights and moving the center of gravity. The Accutron regulators are situated on the inner faces of the magnetic cups, Fig. 121, and the Sonic regulators are at the base of the fork, Fig. 122.

The upper edge of an Accutron regulator is divided into seven equal divisions, each one representing a change in rate of two seconds per day. It is possible therefore to alter the rate of the movement up to twenty-eight seconds per day, but only if the two regulators are in the fully fast or the fully slow position. If both regulators were exactly in the center of their range of movement the amount of regulation would be plus or minus fourteen seconds per day.

The upper edge of the regulators is castellated to provide a means by which they can be moved, and at the same time the castellations serve to indicate each division.

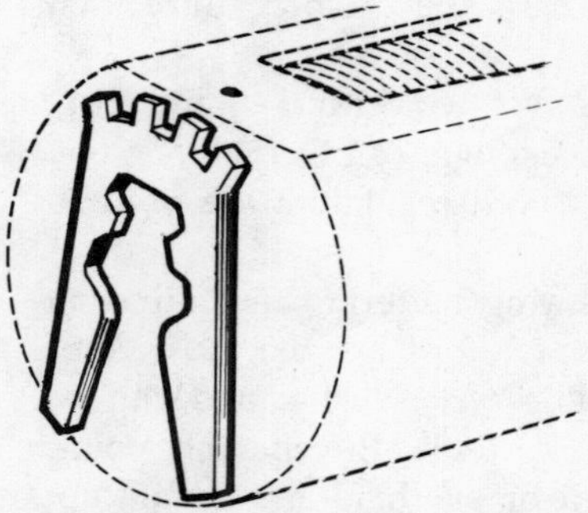

Fig. 121 Accutron regulator

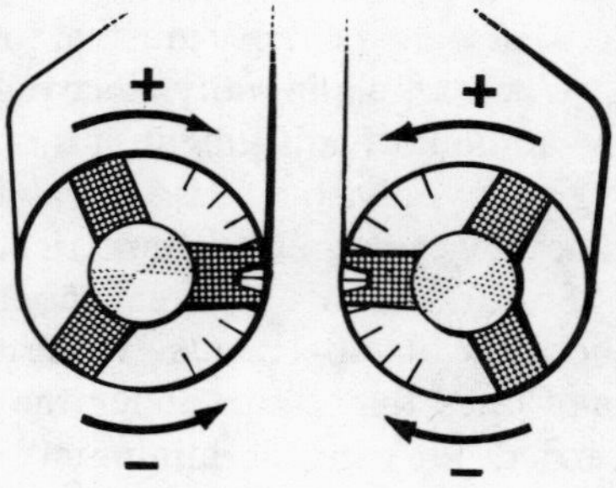

Fig. 122 Sonic regulators

The regulator is pushed around in the required direction by a piece of pegwood stick. On the flat surface of the magnetic cup, adjacent to the shaped edge of the regulator, is a small hole that acts as a datum mark for gauging the distance the regulator is moved.

If the watch is required to run faster, the castellated edge is moved in the direction of the center of the watch. To make the watch run slower the regulators are pushed away from the center.

The regulators of the Sonic watch are turned by the special tool shown in Fig. 123. Each regulator is marked with eight equal di-

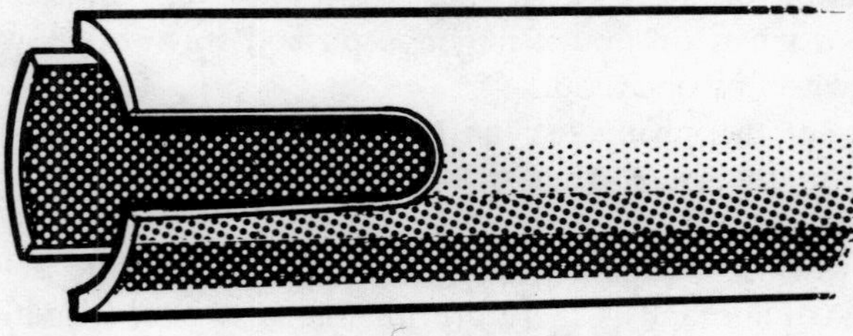

Fig. 123 Sonic regulating tool

visions, where each division represents one second per day. The maximum regulation is therefore sixteen seconds per day, if both regulators are at the fully slow position or the fully fast position. If they are both in the center of their scale, as shown in Fig. 122, then the amount of regulation possible is plus or minus eight seconds per day. The arrows indicate the direction in which the regulators must be turned to achieve a gain or loss.

To regulate the movement of an Eterna Sonic 11T-1550, the movement must be removed from its case and the procedure used for renewing a battery must be followed.

When regulating any electronic watch fitted with a tuning fork, the amount of adjustment, if it is small enough, can be made on one regulator only, but it is desirable that both regulators are used so that they can remain symmetrical.

So far we have spoken about renewing batteries and adjusting the rates of movements. These two operations are the most common ones that come under the heading of servicing, and can be carried out with the minimum of special tools. Beyond this point, any work must be considered as repair or overhaul and should not be attempted without the manufacturers' special tools, and the appropriate electronic tester, complete with a service manual containing operating instructions for the tester.

The Indexing Mechanism—The conversion of vibratory motion into rotary motion is accomplished by a ratchet and pawl device known as the indexing mechanism. Anchored to one of the tines of the fork is a spring index finger, which has a rectangular shaped jewel cemented to its outer end. The jewel engages the teeth of the index wheel and is kept in contact by spring tension.

The theory of the mechanism is to withdraw the index finger with each vibration of the fork so that the jewel drops into the next tooth. Then when the index finger is pushed forward, it will advance the wheel by one tooth.

To prevent the spring tension of the index finger from pulling the wheel in reverse during withdrawal, a spring pawl finger is used.

The pawl finger, which is similar to the index finger, is located on the pillar plate and is therefore unaffected by the vibrations of the fork. While the index finger is advancing the index wheel, the jewel of the pawl finger is riding a tooth. When the index finger completes its forward stroke, the pawl finger drops into the next tooth and holds the wheel steady. The index finger is then free to withdraw, without the possibility of the wheel being disturbed.

In practice this arrangement is not possible because the amplitude or stroke of each vibration of the fork will not necessarily be such as to guarantee that the index finger will be moved just enough to rotate the index wheel one tooth at a time. To make the arrangement a practical one, there must be a means of allowing the tuning fork a flexibility in the amount of movement it can make, and still control the rotation of the wheel to one tooth for each vibration.

The method used in the Accutron movement is illustrated in Figs. 124 to 127.

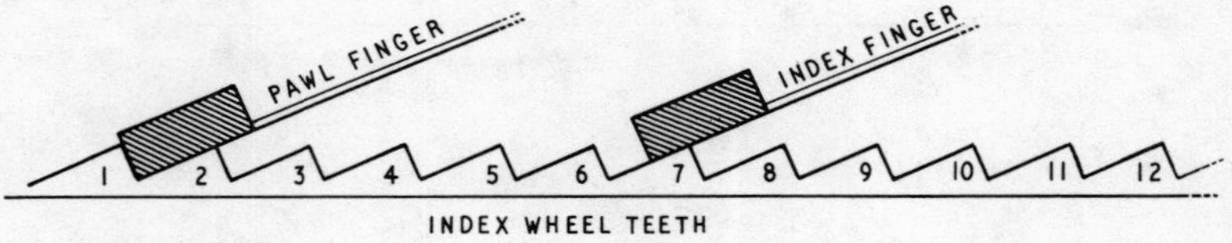

Fig. 124

The index mechanism is so adjusted that when the tuning fork is stationary, Fig. 124, the pawl finger completely engages one of the index wheel teeth, and the index finger rests midway along another tooth. The number of teeth that separate the two fingers is not important.

Let us assume that when the tuning fork is set in motion, the amplitude of each vibration will cause the index finger to oscillate over a distance equal to the length of one tooth, i.e. a half tooth each side of the stationary position.

The index finger will be withdrawn a half tooth, which will cause it to drop into the succeeding tooth, Fig. 125(a). The index finger will then be pushed forward a distance of one tooth, which will turn the index wheel the same amount, Fig. 125(b).

During the withdrawal of the index finger, the pawl finger will hold the wheel steady, and then when the index finger pushes the wheel around, the pawl finger will be lifted by the tooth passing underneath. This will take place until the forward stroke is complete, at which time the pawl finger will drop onto the next tooth.

Now let us consider what will happen if the amplitude of the fork is equivalent to two teeth, i.e. the distance moved by the fork will be equal to one tooth each side of the stationary position, Fig. 126.

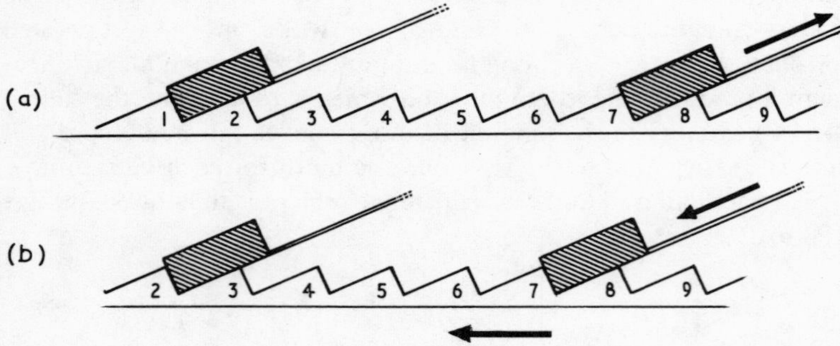

Fig. 125

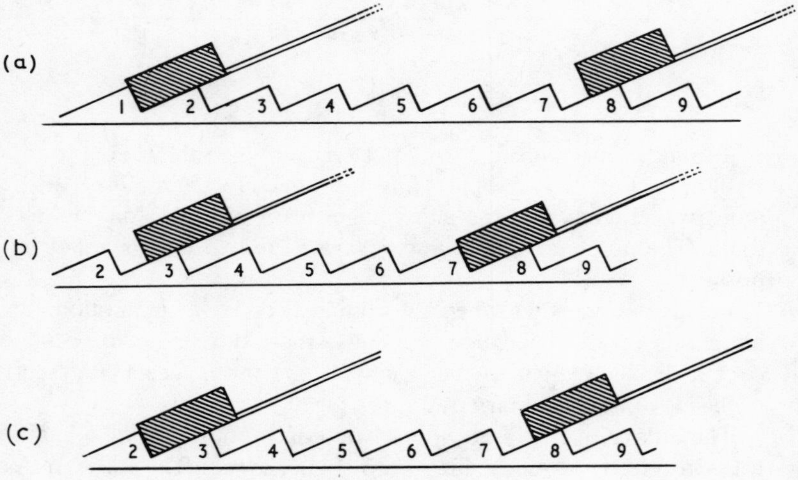

Fig. 126

The withdrawal of the index finger, a distance of one tooth, will cause it to drop from its tooth of rest, and move to the midway position of the succeeding tooth, Fig. 126(a). When the index finger moves forward a distance of two teeth, the first half tooth is made without advancing the wheel. The rest of the forward movement advances the wheel by one and a half teeth, Fig. 126(b). The wheel must rotate beneath the pawl finger because it never moves, and in this case this will bring the pawl finger to rest halfway along the next tooth.

To complete the cycle of vibration, the fork must return to the central position, which causes the index finger to be withdrawn by one tooth. The pull on the wheel caused by the spring tension in the index finger is enough to rotate the wheel backwards. This reversal of direction continues for a half tooth, until the wheel is arrested by the pawl finger, Fig. 126(c). From the illustration it will be seen that once again the wheel has been advanced only one tooth in one vibration.

We will repeat this sequence of movement once more, but this time assume that each vibration is equal to a distance of three teeth, Fig. 127.

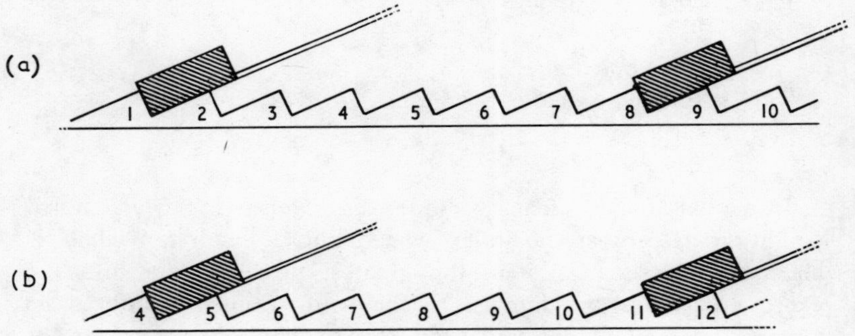

Fig. 127

The first withdrawal of the index finger from the position of rest will be over a distance of one and a half teeth. This will take it from midway along tooth No. 7 in Fig. 124, to a position far enough for the jewel to drop onto tooth No. 9 in Fig. 127(a). The index finger is then moved forward over a distance of three teeth which advances the wheel by the same amount. The wheel is held in this new position by the pawl finger while the index finger is withdrawn.

From these three situations it will be seen that the tuning fork amplitude can vary between a little more than one tooth and slightly less than three teeth and yet the index wheel is always advanced one tooth per vibration.

For this to be possible, the index mechanism has to be adjusted to allow the end of the index finger jewel to rest midway along the length of a tooth when the movement is stationary. This is accomplished by mounting the pawl finger post on a pivoted bridge, the position of which can be altered by a cam, Fig. 119. By altering the position of the pawl finger, the position of the index wheel can be changed in relation to the index finger.

Manufacturers have different methods of adjusting their index mechanisms, and reference must be made to the appropriate service manual.

Cleaning—Ultrasonic cleaning is desirable with electronic watches but the parts mentioned below should not be put into the machine:

(a) electronic circuits
(b) tuning forks
(c) batteries

It is advisable to clean the index mechanism in an assembled condition to protect the index wheel, but if the index wheel is cleaned as a separate item the greatest possible care must be exercised to prevent damage to the teeth. Some manufacturers supply a special holder that supports the wheel by its pivots for this purpose. When handling an index wheel, the tweezers should not grip the rim but the wheel should always be held by its pinion.

Cleaning fluid must be completely free of ferrous metal particles (iron or steel), because they might adhere to the magnets. In any case, close inspection after cleaning is good practice. If metal particles are clinging to the magnets, they can be removed quickly and safely by applying adhesive tape such as Cellotape or Scotch tape.

Over-oiling the bearings can allow oil to get onto the index wheel teeth, which can attract particles of dirt and dust. If this should happen, it may interfere with the correct functioning of the index system.

To repeat what has already been said, because it is well worth emphasizing, a repair or overhaul of an electronic watch should not be attempted without the manufacturers' electronic tester and service manual, and any special tools that may be necessary.

Fault Finding Chart—Tuning Fork Movement

Symptom	Possible Cause	Action
Gains or loses within the manufacturers limits		Regulate
Gains or loses excessively	Bad contact between battery and retaining strap	Clean contacts
	Tuning fork not vibrating freely	Remove movement and check current
	Indexing mechanism out of adjustment	Regulate
	Foreign matter clinging to magnets	Remove foreign matter with transparent adhesive tape
	Mechanical resistance in movement	Disassemble and clean
Hour and minute hands stopped	Hands fouling	Free the hands
Tuning fork hums and sweep second hand turns	Dial train dirty	Clean
All hands stopped but tuning fork hums	Hand set crown not pressed in	Press crown in
	Exhausted battery	Renew battery
	Indexing mechanism out of adjustment or dirty	Clean and adjust
	Movement dirty	Clean
No hum	Exhausted battery	Renew battery
	Movement dirty	Clean the movement
	Faulty electronic circuit	Test and renew faulty component

Part Three

CLOCKS

XIV

Cleaning and Adjusting

When a watch is in need of attention it is usually taken to the watch repairer by the customer. This does not necessarily apply to clocks. A customer may have a particularly heavy mantel clock, such as one made of marble, and no means of transportation.

Sometimes the owners of particularly fine clocks that are fitted with sensitive pendulums would rather a clock repairer visit their homes for a preliminary inspection. If it is found necessary to transfer the clock to a workshop, the repairer is expected to assume the responsibility. Grandfather clocks can present problems and many clock repairers with limited workshop space would much rather remove the movement and leave the case with the owner.

If a clock is gaining or losing time, or has stopped altogether, it does not necessarily mean that it must be taken into a workshop. The cause of the trouble can sometimes be found and rectified without taking the clock out of the customer's home.

A clock repairer must therefore be prepared to make visits, and in some instances to perform services away from his workshop. Quite often, a few touches with the clock oiler will put things right, but if the trouble cannot be rectified within a short time, it is advisable to remove the clock to the workshop where bench facilities exist.

If the clock is fitted with a balance, it can be transported without any preparation other than normal precautions against damage. If, however, the clock is fitted with a pendulum, then the pendulum must be removed and the crutch tied to one side very lightly with a length of sewing thread.

If it is a weight driven movement, the weights must be removed, and it is advisable to rotate the barrels to wind up the gut lines. A few turns of thin string will usually suffice to keep the lines in position on the barrels. A simple but efficient method of carrying a large clock movement is in a cardboard box lined at the bottom with lightly crumpled newspaper, and with tissue paper or crumpled newspaper carefully placed around the movement.

The basic principles involved in the overhaul of a conventional mechanical clock are much the same for all makes and models regardless of whether the movement is driven by weights or spring, or whether it employs a balance or pendulum.

The first step is to carry out a preliminary inspection before any dismantling is done, and to make notes of any points that are in need of attention. Frequently the cause of the trouble can be found at this stage, and corrective action taken without incurring unnecessary expense.

Fig. 128 illustrates a typical medium priced pendulum clock with strike and chime. This particular model, which was made by Smiths of England, is typical of many similar models in regular use today.

If the movement has to be disassembled the work is done systematically, and inspections and any notes are made at each stage. Begin by removing the pendulum and then the hands and the dial. Some hands are made with round holes and others have square holes. Those with round holes are a push fit relying on friction, and those with square holes are usually a loose fit, and are held in place by a flat washer and taper pin, or a round knurled nut. Tight hands should be removed by a pair of hand levers or by a hand removing tool. If levers are used, the dial must be protected by a piece of chamois leather or folded cloth.

The method of securing the dial will also vary with the type of clock. In many instances, the dial is screwed to the case and when the hands have been removed the movement is withdrawn from the rear, leaving the dial undisturbed on the case. With some

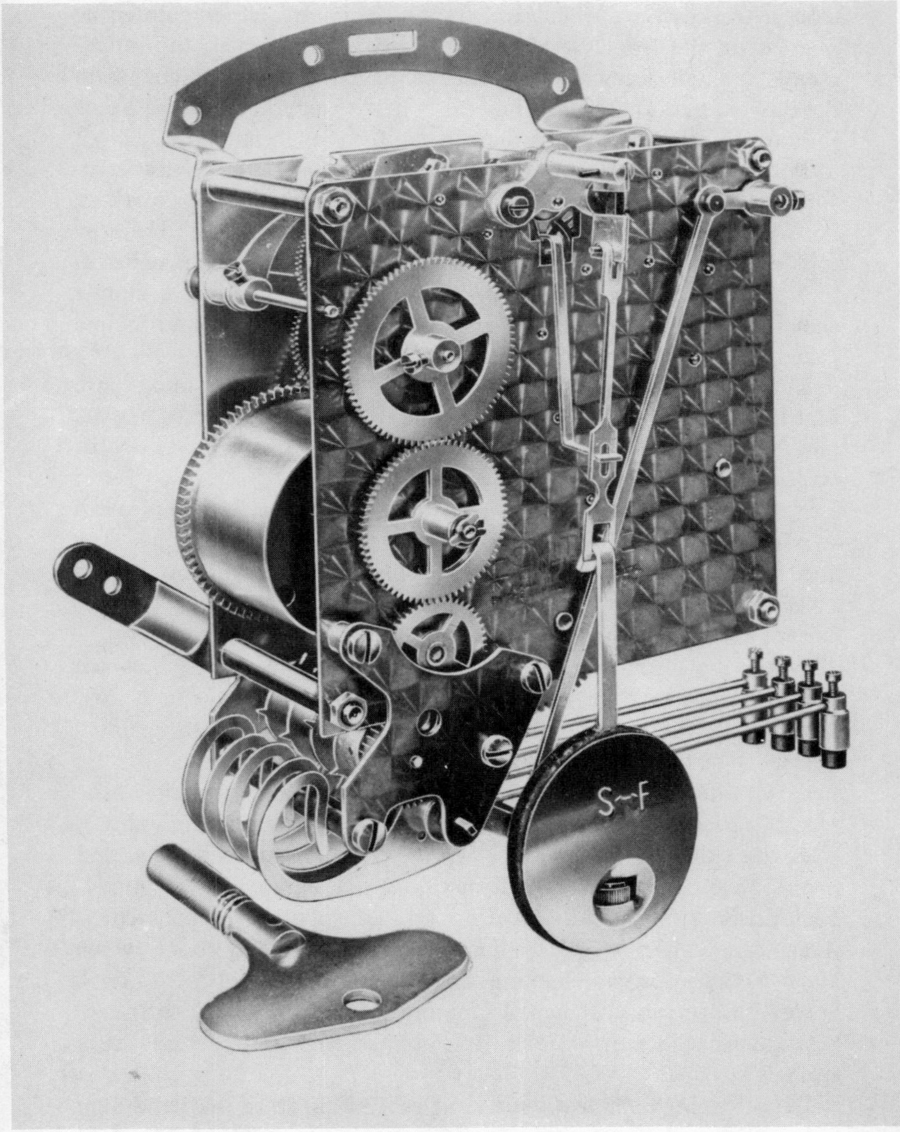

Fig. 128 English pendulum with strike and chime

clocks, the case is the dial and the numerals are either painted or planted on the front. Sometimes the dial is secured to the movement by small screws. Other clocks have dials with feet. When the dial is fitted to the movement it is held in position by small pins that are inserted in holes drilled in the feet.

If the layout of the movement is unfamiliar, it is advisable to study the strike and chime mechanisms until a good working knowledge of their action has been obtained. This is best done by replacing the minute hand and using it to set the mechanisms in motion. Here again is another situation when making a simple sketch showing the relative positions of the strike and chime parts may save a lot of time at the assembly stage.

Turn the movement through a twelve hour cycle to make sure that the chimes and strikes are functioning at the correct moments and in the right sequence. It is better to be aware of faults at this stage than to find them after assembly.

Note where on the snail the pin of the strike rack falls prior to striking the hour. The pin should drop into the corner of a step and not anywhere along the step. This can be regulated during assembly by gearing the hour wheel with the minute wheel in the correct position by trial and error.

When the heart-shaped gathering pallet rotates, the pallet pin should engage the rack teeth cleanly and with plenty of depth. Adjustment can be made by carefully bending the pin of the gathering pallet in the required direction.

Check the action of the escapement by watching the drop of the escape wheel teeth onto the pallet while the crutch is moved slowly from one side to the other. This type of clock is usually fitted with a recoil escapement, so named because the escape wheel moves backwards after a tooth has dropped. This type of escapement is fitted with either a solid pallet or a strip pallet, both of which allow a wide range of locking with the escape wheel teeth. It is, however, necessary that the amount of locking be controlled, because trouble can arise if it is excessively deep or not deep enough.

Deep locking calls for a wide angle of swing from the pendulum before the tooth at the exit pallet can be released, or the incoming tooth can be admitted. This increase of swing demands more power to impel the pendulum. If the depth of locking is such that the pendulum is unable to swing the pallet enough to clear an outgoing or incoming tooth, then the movement ceases to function.

By contrast, if the locking is too shallow, there is the possibility that mislocking will take place, which can result in an erratic gain in timekeeping.

The pallet arbor pivots must be checked for sideshake in the pivot holes and if the amount of shake is too much, then the holes will have to be bushed, and possibly the pivots will require some attention between centers in the clockmakers throw. These pallets are provided with a means of adjustment, but until excessive wear between pivot and pivot hole has been eliminated, the question of adjustment to provide good depthing cannot be considered.

The mainsprings should now be let down. Place the movement, front plate on top, on the edges of a stout cardboard box for support. Position the winding key on the squared end of the winding arbor and turn in the direction of winding. Hold the pawl clear of the ratchet wheel with a piece of pegwood, and allow the mainspring to unwind under the control of hand pressure. Do not relax your grip on the key because the sudden release of the mainspring could do damage to your fingers and might easily damage the clock. As a precaution it is wise to hold the pawl in such a way that it can be instantly released in case the key slips between the fingers. When the pawl is held away from the ratchet wheel the pressure of the mainspring will cause the key to turn backwards. When you are no longer able to turn your hand at the wrist, the pawl is released into the ratchet wheel which allows your hand to reposition itself. Repeat this procedure until the spring is fully unwound. Treat each of the three mainsprings in the same way and when the power has been shut off it will be safe to disassemble the movement.

First, we must check the wheel trains for depthing and all pivots for sideshake. Place the movement on the cardboard box, back plate face up, and remove the back cock and the pallet arbor assembly. The wheels will now be free to spin. Hold the escape wheel steady by finger pressure, and rock the third wheel backwards and forwards to obtain an impression of the depthing between the teeth of the third wheel and the leaves of the escape pinion. Then apply some finger pressure to the side of the escape wheel arbor to increase its resistance to turn and, with a finger of the other hand, turn the third wheel round a few revolutions. The sensation should be smooth, and any signs of roughness or unevenness must be noted for correction when the movement has been disassembled. Continue this procedure throughout the train until all wheels have been checked for depthing and notes made of those wheels and pinions that require further investigation.

The sideshake of all pivots must now be checked in the same way as the pallet arbor pivots, and notes must be made of those pivot holes that require bushing.

Leave the movement on the cardboard box and remove the motion work, the rack with its associated strike pieces, the release piece, the strike and chime flirts, and the remaining chime pieces on the front plate. Turn the movement over and remove the hammer work, the chime pin barrel, and the strike operating lever.

There are different methods of securing strike and chime pieces to their arbors. Wheels and racks are often held in place by flat washers, followed by spring steel circlips that snap into position in grooves cut close to the ends of the arbors. The steel flirts are usually riveted to brass collars, which are a drive fit onto their arbors. A hole is drilled through the arbor in such a way that when the movement is assembled, the plate is positioned behind the collar, but in front of the hole. To prevent the arbor from being withdrawn, a short length of soft steel wire is passed through the hole and each end is bent over. When the movement is reassembled, new lengths of wire are used. A brief inspection of the movement will soon disclose what methods are used to secure the strike and chime pieces.

The front and back plates are held apart by a shouldered post at each corner. These posts are either threaded at the ends to take nuts or the posts are drilled for taper pins. In any case, one end of each post must be freed in order to lift off the front plate. Do not force it; at the least sign of *any* resistance, inspect the movement to find the cause.

When the front plate is removed, the strike gathering pallet and gathering wheel, the center wheel and cannon pinion, and the chime locking plate and cam wheel will be with it.

If the center wheel arbor hole can be properly cleaned, there is little or nothing to be gained by further dismantling. If it is necessary to remove the cannon pinion, this is simply done by holding the plate in the palm of the hand and striking the end of the center wheel arbor with a brass hammer. The strike gathering pallet and the chime locking plate are both removed in the same way.

With the front plate off, the barrels and the wheels can be lifted out. With a sharp pointed instrument mark the three barrels with a small letter G, S or C to indicate 'going,' 'strike' and 'chime.' The two flys should also be marked S and C. In this way we can be sure that the pieces concerned will be refitted correctly at the assembly stage.

The covers of the mainspring barrels have a crescent shaped cut-away. Insert the blade of a screwdriver and gently raise the cover from the barrel. Grip the arbor in a pair of brass faced pliers, and turn it backwards until it can be unhooked from the inner end of the spring, and then lift it out. If the spring and the interior of the barrel are clean and adequately lubricated it is better to leave them as they are. Put them to one side in a clean tin with a lid to keep out dirt.

If the barrel is dirty inside and the oil is black, then the spring must be removed. This operation is easily done but care is needed. Hold the barrel in one hand in such a way that the thumb passes across the coils to prevent them jumping out. With the other hand, hold the inner end of the spring in the jaws of the brass faced pliers, and slowly and gently ease out the first two or three coils. Discard the pliers and continue easing out the spring with your fingers. Providing a good grip on the barrel is maintained and the coils are prevented from springing out, there will be no difficulty. When the outer end of the spring has been reached, it will be necessary to push it backwards to unhook it from the barrel wall.

We must now refer to our notes to see which pivot holes, if any, were in need of bushing, and then inspect the condition of the pivots that operate in these worn holes. Most probably we will find that they too are worn, some with ridges, and others with a burr on the end. The repair procedure detailed in Chapter XV must be followed until all worn pivots and holes have been remedied. After that, we will be ready for cleaning.

There are two methods of cleaning a movement; one is by means of a machine, and the other is by hand. The mechanical method is explained in Chapter VI and so in this chapter we will talk about the hand method.

A round tin about 6 in. in diameter and approximately 3 in. deep will be required for use as the cleaning bath. An ordinary ½ in. paint brush is suitable for cleaning, providing the brush is new and has not been used for some other purpose. We will need some pegwood sticks, a piece of clean soft cloth such as an old handkerchief, a long narrow tapered strip of chamois leather, a newspaper, and some gasoline.

Pour the gasoline into the tin to a depth of about 1 in., and hold the parts over the fluid. Brush each piece separately, and continually dip the brush into the gasoline. Take care to remove every trace of foreign matter, and make sure that none of the old dried oil remains.

Meticulously brush between the wheel teeth and dab the bristles of the brush into the leaves of the pinions. With a razor blade, shape the end of a pegwood stick rather like a twist drill, so that it fits into the oil sinks. It should remove any stubborn pieces of dirt by twirling it between finger and thumb.

Small pivot holes can be cleaned by pegging. Insert the tapered end of another pegwood stick into the pivot holes and rotate it. Remove the stick, scrape away the dirt, and then repeat the process.

The large holes are cleaned with the chamois leather. Clamp the wide end of the strip in the jaws of the vise and thread the narrow end through the hole. Hold the narrow end in one hand and pass the plate up and down the leather with the other hand.

After washing each part, put it to one side on several thicknesses of newspaper to drain, and then dry the parts with the cloth.

Wash the mainsprings very thoroughly and make equally certain they are properly dried. Take a small piece of cloth and, starting from the center, wipe along the length of the springs until the outer ends have been reached. The springs can then be refitted into their barrels.

If the springs are not too powerful, they can be coiled into their barrels by hand. This is done by first hooking the outer end of the spring onto the barrel hook, and then turning the barrel and the spring together in your hands. At the same time, use the fingers to manipulate the spring into position.

After the spring has been coiled into the barrel it is necessary to level the spring flush with the barrel floor. Do not tap it down with a hammer or any other tool, and do not attempt to push or lever any of the coils. Any of these methods could easily damage the spring. A slip with a sharp tool such as a screwdriver would most probably score the face of the spring, and weaken it at that point. Under tension the spring could easily break where the score has been made. A burr could be thrown up which would prevent the neighboring coil from sliding smoothly when running down from the wound position. Hammering the edges of the spring could bruise the metal and distort the shape, and affect the coils on either side when fully wound.

The correct method of knocking down the spring is to bring the face of the barrel down smartly onto the bench top two or three times so that the coils will adapt a level and natural position without harm.

If the springs are too strong to be hand wound, then a mainspring winder will have to be used. There are two types of winders, both of which are held in the bench vise. They consist of a housing through which passes a spindle with a handle at one end, and facilities to wind the spring at the other end.

The more simple of the two designs has a hook near the end of the spindle, to which the center of the spring is attached. Turn the handle with the right hand and allow the spring to coil in the palm of the left hand. Hold the spring in the piece of linen during the winding operation.

When winding is complete, release the grip on the handle; it will not rotate backwards under spring tension because a ratchet is fitted to the tool to prevent this. Carefully remove the left hand from the spring, replace it with the right hand and hold it tightly to the spring. With the left hand, bring the barrel to the coiled spring, and place it over the spring as far as it will go. Then remove the linen and slowly allow the spring to unwind into the barrel. When the outer end of the spring has attached itself to the barrel hook, the barrel will want to rotate. Allow the barrel to revolve slowly until all tension has gone from the spring, and then unhook the inner end from the winder spindle.

The other design of mainspring winder is more sophisticated and more expensive. It is supplied in a box with a number of different sized drums. A drum is selected that will enter into the barrel, and then it is fitted to the end of the winder spindle with the open end facing outward.

Inside the drum is a center pillar with a hook, to which is attached the inner end of the spring. The handle is turned and the spring is fed sideways into the drum through the open end, until it is completely coiled. The barrel is then placed over the drum, and by pressing on the handle, the spring is transferred from the drum into the barrel. This method is preferred, and it will wind the strongest springs with ease.

Now that the mainsprings are in their barrels and correctly bedded down, we can insert the barrel arbors. Make sure that the inner end of each spring is sufficiently coiled to allow the spring to be hooked to the arbor. If necessary, reshape the inner end of the spring by reducing the diameter of the innermost coil.

Place a few touches of clock oil on the edges of each spring and a little on the bottom of each barrel, and then snap the covers into position, making sure that they are level with the edge of the barrel. The arbors will then need a touch of oil where they emerge from the barrels and the covers.

Now that all parts have been cleaned, we can start to assemble and oil the movement. Place the backplate on the cardboard box, front face up, and place the three barrels, the wheels, and the flys in position. It may be necessary with some of the arbors to guide the pivots into their holes with tweezers.

When all the back pivots are in their holes, lower the front plate until it is resting on the front pivots. Hold the two plates together with slight finger pressure and guide the front pivots and barrel arbors into their holes in the front plate. When all the pivots are in, the two plates should be resting on the shoulders of the corner posts. The nuts can then be screwed on and tightened, or the taper pins can be inserted and tapped home tight.

Apply just a touch of clock oil to the pivots in both plates and to the three mainspring arbors. Refit the ratchets and wind the going mainspring two or three clicks; if the pivots are free, the going wheels will spin.

If the cannon pinion was removed it should now be refitted. Support the backplate on a flat block of wood and tap the cannon pinion onto the center arbor using a hollow punch. Make frequent checks to ensure that the cannon pinion is not driven up to the front plate; slight endshake must be left for the center arbor.

Give a touch of oil to the minute wheel post and then the motion work can be refitted.

Turn the movement over, locate the front pivot of the pallet arbor in the front plate, and lower the back cock into position. The back cock carries the rear pivot of the pallet arbor, and by altering the position of the back cock, the depth of locking between the pallet faces and the escape wheel teeth can be adjusted.

Clocks that are fitted with solid pallets have an adjusting screw above the back cock. By slackening the back cock screws and turning the adjusting screw clockwise, the back cock is raised and the depth of locking is reduced. Turning the screw counter-clockwise will have the reverse affect. When adjustment is complete the back cock screws are tightened.

Adjustment for clocks fitted with a strip pallet is provided by means of elongated screw holes in the back cock. The screws are slackened, the position of the back cock is altered and the screws are retightened.

The strike and chime pieces can now be fitted, using the sketch that was made before the movement was disassembled as your guide.

A touch of oil will be needed for the arbor pivots and for all sliding faces, but do not over-oil.

Check the strike and chime mechanisms by partly winding the mainsprings and turning the movement by using the minute hand. If the sequences are correct, the movement can be refitted to its case.

XV

Bushing

We have already seen in Chapter V that the depth of mesh in gearing is important if good rating and timekeeping is to be achieved. Watch and clock designers most carefully calculate the diameters of wheels and pinions, and the distances the pivot holes are from each other, in order to obtain correct depthing.

If a movement is allowed to run dry for any length of time the arbor pivots and the pivot holes begin to wear. Minute particles of steel from the pivots embed themselves in the walls of the brass pivot holes and act as an abrasive, causing the pivots to wear at a greater rate than the holes themselves. The shape of a badly worn pivot is usually ridged because the steel particles that are doing the damage are irregularly spaced. The holes themselves increase in diameter and usually wear an oval where the side loading is greatest.

Immediately as pivots and pivot holes begin to wear, the side shake is increased, and the depthing of the gears is altered. Wheels and pinions are either too deep or too shallow, depending on which side of the hole the pivot is operating.

It is not possible to explain when the side shake is excessive and when it is not, because much depends on the type and size of movement. To make a comparison with a new or nearly new movement of similar design would be a very good method of assessment, but the opportunity of doing this does not often present itself. It is something that one gains with experience.

Sometimes one can hold the wheel between the finger and thumb, and move the arbor from side to side. Other times, when the arbor is more inaccessible, it has to be held between strong tweezers and moved. With either method, one has to assess the amount of movement. It is sufficient to say that if the pivot and the pivot hole were clean and dry, one should be able to just feel the freedom. Anything in excess of this means the pivot and its hole should be the subject of further inspection after disassembly.

Weight driven clocks are prone to this defect because long after the oil has dried from the bearings, the weights have sufficient energy to overcome any resistance to turn that may exist in the gear train. Consequently, a high rate of wear can develop if these conditions are allowed to continue.

Fortunately it is not a common occurrence but sometimes clocks are brought into workshops that are so filthy they have to be dismantled, cleaned and reassembled before they can be inspected.

Both pivots of each arbor must be checked in their holes in two directions, and notes made listing the pivots and pivot holes that need attention. The pivots are repaired first, and then the pivot holes are rebushed and brooched to suit the pivot. It is unusual for a pivot to be so badly worn that only a new arbor will put matters right. The more usual repair is to remove the ridges and then finish with a burnisher.

In Chapter III we discussed the many uses of a watch repairer's turns. A clockmaker has a similar but larger bench tool known as a throw. Its functions and methods of application are almost identical to those of the turns, but instead of being operated by a bow it is powered by a large diameter, hand-operated wheel and gut string. The obvious difference is that with the turns the direction of rotation alternates, whereas with the throw the direction remains constant.

An arbor with a worn pivot is fitted up in the throw between dead centers, and with a curved graver the ridges are turned off. Some repairers prefer to use a very worn, dead smooth file, but it is my opinion that the graver is the more accurate.

Having removed all traces of wear, the arbor is fitted up with a dead center at one end, and a grooved pivot runner at the other end, and the pivot is burnished as described in Chapter III.

When all the worn pivots have been restored to a new condition, their respective pivot holes in the plates will have to be bushed. There are three methods of doing this, each employing a different type of bush. They are known as the riveted bush, the friction bush and the French bouchons. Let us first consider the riveted bush.

Turn the plate over with the oil sink facing downward, and insert a cutting broach into the pivot hole. Rotate the broach, making quite sure it is held at right angles to the plate. After a few cuts hold the plate up with the broach in position and check for squareness. If the broach is leaning to one side correct the fault at the next cut. Apply just sufficient pressure on the broach to enable it to make light cuts to ensure a true round hole. Too much pressure will cause the cutting edges to dig into the plate and more metal will have to be removed than need be.

Continue the cutting until the hole is round and the walls are of sufficient thickness to support a bush, Fig. 129 (a). When this has been achieved turn the plate over, and with a bevelling tool cut a bevel in the bottom of the oil sink, into which the end of the bush can be riveted, Fig. 129 (b).

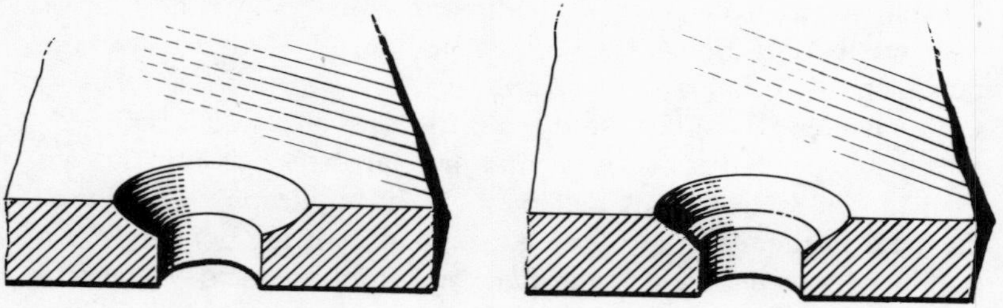

(a)

(b)

Fig. 129 Preparing hole for riveted bush

There are two types of broaches available, both of which are slightly tapered. One is round and is used for polishing the insides of holes, and the other is five-sided, and the edges thus formed are used for cutting and enlarging holes. Broaches are obtainable from tool dealers, and are usually supplied in sets with six or twelve different sizes in tubes.

We now need a length of brass bushing wire. This is hollow seamless wire supplied in lengths of about 3 in. The wire is manufactured in a range of both outside diameters and hole diameters.

Measure the diameter of the pivot with a micrometer, and select a length of bushing wire with a hole slightly smaller than the pivot. Cut a piece of bushing wire a little longer than the thickness of the plate into which it is to be fitted. The bushing wire is placed on a turning arbor, and fitted between centers in the turns or on a lathe. The diameter of the bush is then turned with a very slight taper that is as near as possible to that of the cutting broach used for opening the hole in the plate.

Continue to turn the bush, occasionally removing it from the arbor to try the fit in the hole. The small diameter of the bush should be slightly smaller than the mouth of the hole on the inside of the plate, and the bush should be turned until it can enter the hole a distance of about half its own length. When the diameter is correct, turn the ends square, and at the same time reduce the length of the bush to a little more than the depth of the hole.

Turn the plate over with the oil sink facing downward, and rest it on a riveting or punching stake. Place the bush in the hole, making sure the small diameter enters first, and drive it home with a flat bottom punch until it is flush with the plate, Fig. 130 (a).

With a bevelling tool, cut a bevel in the mouth of the bush until the bottom edge of the bevel is level with the original base of the oil sink. This will enable the oil sink to be restored to its original depth.

Select a round end punch, a little larger in diameter than the bush, and punch the end of the bush until the metal has been spread into the bevel at the base of the oil sink, Fig. 130 (b).

With the bush firmly in postion in the plate, the hole must be opened to receive the pivot. Select a cutting broach of suitable size, and enter it from the inside of the plate. Twirl the broach between finger and thumb, and apply only light pressure to avoid the possibility of loosening the bush.

Check to be sure the broach is perpendicular to the plate and correct any fault, as previously described. Make frequent trials by offering the pivot to the hole, and stop cutting as soon as it is possible to enter the pivot with a tight fit.

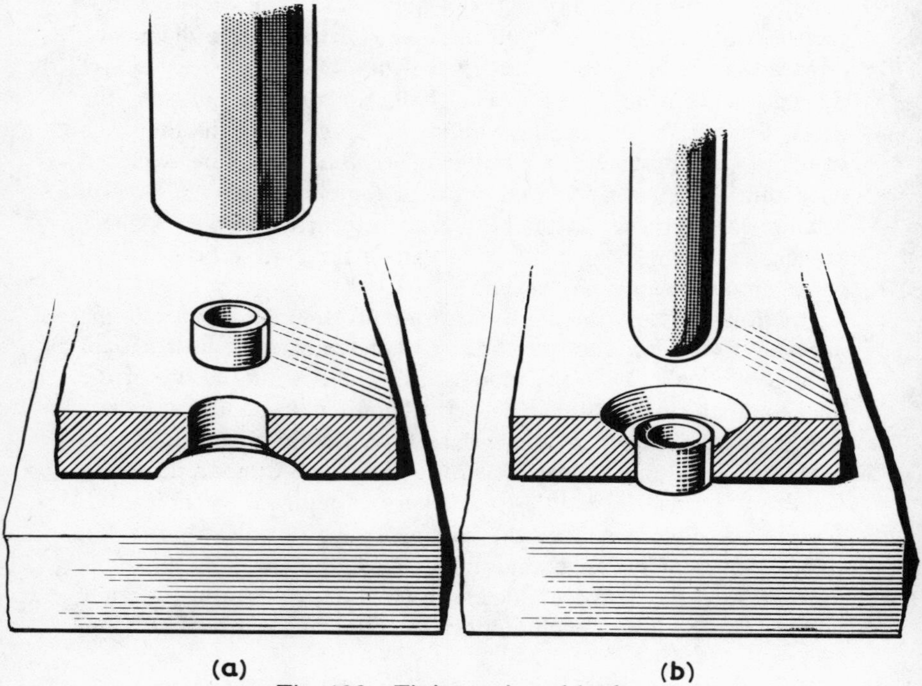

Fig. 130 Fitting a riveted bush

The hole is finished by burnishing with a round broach. Select a suitable size broach and clean it by rubbing along its length with a piece of very smooth emery paper. Keep the grain straight and do not rotate the broach during cleaning.

Apply a film of clock oil to the broach, and insert it in the bush from the inside of the plate. Spin it lightly and quickly between your fingers. This will cause the surface skin of the hole to be compressed, enlarging the diameter and producing a hard polished finish. Occasionally check the fit of the pivot and stop burnishing when the pivot enters freely with just the right amount of side shake.

After the bush has been riveted by the round end punch, it may be necessary to skim the oil sink. This operation is carried out by using a suitable size chamfering tool or a rose cutter. Rose cutters are available in sets, some of which include detachable pilot pins. A pin to fit the pivot hole is inserted in the end of the appropriate cutter, and when the cutter is in use the pin will hold it concentric to the hole.

All that now remains to complete the task is to clean up the inner end of the bush flush with the plate, and at the same time remove any burr that may have been caused by the broach. This is best done by a small, flat, dead smooth file. Place the extreme end of the file on the bush and hold the same end of the file by its edges between finger and thumb. Move the file backwards and forwards over a distance of about ⅛ in., and at the same time slowly turn the plate. This will prevent any tendency of the file to wear a shallow groove in the plate, but with a little care it will be possible to dress back the bush, leaving hardly a mark on the plate.

The next method of bushing that we are going to discuss is the friction bush system. This is a more modern approach involving the use of a universal bushing tool, but for those who have to bush many holes it is a better method because it is quicker, and accurate results are more easily attained.

The universal clock bushing tools produced by Bergeon and by KWM are very good. Universal tools consist of a base and vertical headstock assembly. The base is used as a staking tool, and supports a bed on each side. Two adjustable clamps are mounted on these, one on each side of the base.

Accessories are available which include anvils, reamers, pushers, chamfering cutters, pivot gauge, and ready made bushings in a range of different sizes.

In addition to the universal tools, there are other bushing aids which are less sophisticated and are in a lower price range. Details of these can be obtained from your tool dealer.

The plate to be bushed is placed horizontally with the oil sink downward, and is loosely carried in the clamps. A male centering tool is placed into the vertical headstock. The centering tool is lowered by means of the headstock screw, and the position of the plate is adjusted until the point of the tool enters the hole to be bushed. Hold the centering tool down as far as it will go and tighten the two clamps onto the plate. Remove the centering tool, and the plate is ready for the hole to be reamed.

Friction bushes are supplied in a range of sizes and in a finished condition. The lid of each container is marked to show the height, the bore, and the diameter of its content. All measurements are in millimetres, and it will therefore be necessary to measure the diameter of the pivot with either a metric micrometer or a pivot gauge supplied for this purpose.

Select a suitable bush and note the size of the diameter. Then choose a reamer, the diameter of which is three hundredths of a millimetre smaller, i.e. $3/100$ mm or 0.030 mm. This is very important because the success of the bushing operation relies almost entirely on the correct choice of the reamer for the particular bush to be inserted. It is this difference of 0.030 mm in the two diameters that provides the friction tightness that holds the bush firmly in the plate.

If, for example, the diameter of the pivot was measured and found to be 1.48 mm, a suitable bush would be one with a bore of 1.60 mm. See the chart at the end of the chapter. If the diameter of this bush was indicated to be 2.70 mm then the reamer would have to be 2.67 mm.

A further consideration to be taken into account when choosing a bush is the height. The chosen bush should be of such a height that when it is pressed into position flush with the inside face of the plate, the chamfered end of the bush is level with the bottom of the oil sink in the plate. This is not a dimension that can easily be measured. The more practical method is to estimate the distance.

Fit the reamer in the headstock and rotate the wheel at the top, bringing the reamer down onto the plate. Let the reamer cut its way into the hole naturally and without any force. Continue with the cutting until the full length of the cutting edges have passed right through the plate. When the round shank of the reamer enters the plate, the diameter of the hole will be correct.

At each end of the hole the reamer will throw up a small burr which must be removed. Take out the reamer and fit the chamfering cutter in the headstock. Remove the burr from the inside of the plate, and then release the plate and turn it over. It will have to be centralized with the centering tool as before and retightened in the clamps. Refit the chamfering cutter, remove the burr from the oil sink end of the hole, and then take out the chamfering cutter. The plate must now be turned over with the oil sink facing downward, repositioned, and clamped as before.

Fit the pusher into the headstock, and position the bush in the hole with the flat end uppermost. Drive the bush into the plate by tapping the pusher holder with a mallet. Tap the bush right home flush with the plate.

Try the pivot for fit in the bush hole. Sometimes the holes close up slightly while being pressed into the plate, and they require opening out a little. This is done as previously described with a cutting broach, and is finished with a round broach. Here again it is necessary to take the precaution of not forcing the cutting broach so that the friction fit is not disturbed.

The chart shows a range of KWM bushings. To select the correct bushing, first measure the pivot in millimetres, locate this dimension in the first column of the table, and then read across. Where more than one size is listed, select the height that is required from line H.

When the selection has been made, the diameter of the bush can be seen in line D, and the corresponding reamer size is given in line R. The reamers themselves are identified by roman numerals as shown in line RN. The numbers prefixed by the letter L are the catalog or reference numbers for ordering.

Fig. 131 French bouchon

The third method of bushing is the French bouchon, Fig. 131. These bushes, or bouchons as they are known, are lengths of hollow brass wire about 20 mm long, which have been turned with a

PIVOT DIAMETER	BORE																
		H	1.0			1.4			1.7		1.9			2.7			3.0	4.0
		D	1.2	1.8	1.8	2.7	3.5	5.9	5.9	1.8	2.7	3.5	1.8	2.7	3.5	5.9	8.7	8.7
		R	1.17	1.77	1.77	2.67	3.46	5.8	5.8	1.77	2.67	3.46	1.77	2.67	3.46	5.8	8.6	8.6
		RN	1	11	11	111	1V	V	V	11	111	1V	11	111	1V	V	V1	V1
0.07–0.08	0.10		L55															
0.10–0.12	0.15		L56															
0.15–0.17	0.20		L01															
0.19–0.21	0.25		L57															
0.23–0.25	0.30		L02		L64					L86								
0.27–0.30	0.35		L58															
0.32–0.35	0.40		L03		L08					L87								
0.38–0.40	0.45		L59															
0.42–0.45	0.50		L04		L09					L32			L94					
0.48–0.50	0.55		L60															
0.53–0.55	0.60		L05		L10					L33			L95					
0.57–0.60	0.70		L06		L11					L34			L96					
0.65–0.70	0.80		L07		L12					L35			L97					
0.75–0.80	0.90			L61	L13					L36			L98					
0.85–0.90	1.00			L62	L65	L14				L37				L99				
0.95–1.00	1.10			L63	L66	L15					L38			L100				
1.05–1.10	1.20					L16					L39			L101				
1.15–1.20	1.30					L17					L40			L102				
1.25–1.30	1.40					L18					L41			L103				
1.35–1.40	1.50					L19					L42			L104				
1.45–1.50	1.60					L20					L43			L105				

1.55–1.60	1.70	L67				L88		L106			
1.65–1.70	1.80	L21				L44		L107			
1.75–1.80	1.90	L68				L89		L108			
1.85–1.90	2.00		L22				L45		L109		
1.95–2.00	2.10		L69				L90		L110		
2.05–2.10	2.20		L23				L46		L111		
2.15–2.20	2.30		L70				L91		L112		
2.25–2.30	2.40		L24				L47		L113		
2.35–2.40	2.50		L71				L92		L114		
2.45–2.50	2.60		L25				L48		L115		
2.55–2.60	2.70		L72				L93		L116		
2.65–2.70	2.80		L26		L79		L49		L117		
2.75–2.80	2.90			L73	L80						
2.85–2.90	3.00			L27	L50					L118	
2.95–3.05	3.20			L28	L51					L119	
3.10–3.25	3.40			L74	L81					L120	
3.30–3.45	3.60			L29	L52					L121	
3.55–3.65	3.80			L75	L82					L122	
3.75–3.85	4.00			L76	L83					L123	
3.90–4.05	4.20			L30	L53					L124	
4.15–4.25	4.40			L77	L84					L125	
4.35–4.45	4.60			L31	L54					L126	
4.55–4.65	4.80			L78	L85					L127	
4.75–4.85	5.00										L128
5.00–5.05	5.20										L129
5.15–5.25	5.40										L130
5.35–5.45	5.60										L131
5.55–5.65	5.80										L132

PIVOT DIAMETER	BORE																	
		H	1.0		1.4				1.7	1.9			2.7				3.0	4.0
		D	1.2	1.8	1.8	2.7	3.5	5.9	5.9	1.8	2.7	3.5	1.8	2.7	3.5	5.9	8.7	8.7
		R	1.17	1.77	1.77	2.67	3.46	5.8	5.8	1.77	2.67	3.46	1.77	2.67	3.46	5.8	8.6	8.6
		RN	1	11	11	111	1V	V	V	11	111	1V	11	111	1V	V	V1	V1
5.75–5.85	6.00																L133	
5.95–6.05	6.20																	L134
6.15–6.25	6.40																	L135
6.35–6.45	6.60																	L136
6.55–6.65	6.80																	L137

H = height
D = diameter
R = reamer size
RN = reamer identification

Chart of KWM Bushings

very slight taper. They are supplied in different diameters and with different sizes of holes.

Select a bush with a hole suitable for the pivot, and then choose a cutting broach slightly smaller than the diameter of the bush. Turn the plate over, oil sink facing downward, and broach the hole as previously described for the riveted bush. The diameter of the hole has to be increased until the tapered end of the bouchon will just enter. Hold the bouchon in a pin vise and push it into the hole. This should be a tight fit. The circumference of a bouchon is slightly undercut at a convenient distance from the end and so, once it is held firmly in the plate, it can easily be snapped off.

The new bush is now filed flush with the inner face of the plate, the oil sink end is riveted, and the pivot hole is opened slightly with a cutting broach. It is finished with a round broach in exactly the same manner as was used in the riveted bush method.

XVI

Turret Clocks

THERE ARE STILL MANY MECHANICALLY OPERATED TURRET CLOCKS in working order in public buildings and churches. The oldest in the United States is in Whitfield House, Guilford, Connecticut, and was built in 1726. Unfortunately there are fewer men who are prepared to undertake the maintenance of these old clocks. The work is heavy and dirty, and sometimes involves climbing dark narrow stairs to work in uncomfortable and drafty conditions. Enough, you might say, to discourage the most agile and stout-hearted of clock repairers. Maybe so, but for the man who wishes to earn for himself a few dollars, a 'winding contract,' as they are often called, might be very acceptable. If the local population came to rely on the accuracy of the clock it could be useful publicity for the repairer, if he was engaged in general horological work.

Negotiations for a contract usually start by approaching local authorities and church officials with a request that you be considered for the work when next there is a vacancy.

The tools that are needed are those of a general engineer and a suggested kit is as follows:

1 8 oz. hammer
1 pair side cutting pliers
1 8 in. medium file

1 wheel brace
assorted twist drills
1 ⅛ in. pin punch
1 oil can
1 battery lamp
1 9 in. ratchet screwdriver
1 12 in. adjustable wrench
1 1 in. paint brush

The terms of a contract or agreement are simple enough. One is expected to make regular visits, say once each week, to wind the clock, carry out an inspection, and make any repairs that may be necessary. Adjustments to maintain accurate timekeeping would be made as required.

When a clock repairer is hired for this work, it is usual for him to understudy the man who has been looking after the clock. Two or three visits are all that are necessary to learn the best methods of handling the various routine tasks. One would also need to know the names and addresses of turret clockmakers who had supplied materials in the past, and the names of engineers, welders, and machinists, who would undertake the repair or manufacture of a broken or worn piece.

It is always good to start with a general examination and make notes. Make sure the movement is clean, well-oiled, and protected from the weather. Cleaning is done by brushing with kerosene and wiping dry with a clean cloth. Bearings are lubricated with turret oil obtained from a turret clockmaker. Rust is removed by rubbing with an emery cloth and machine oil, or a scraper or wire brush if it is severely rusted.

Clean the bare metal by brushing on some gasoline, and then paint it with a galvanizing paint. When dry, any paint appropriate for external use may be applied.

New metal parts, if they are iron or steel, should be painted with red oxide primer to prevent rust from forming, and then given a top coat of enamel paint.

Keep a watchful eye on the steel lines, because at the first sign of fraying a new line should be ordered. A new line is best rubbed with tallow before fitting to prevent it from rusting. When fitting the line, take care not to introduce any kinks.

During winding, the weights are out of sight in many turret clocks and because most of them are made without stop work, there is a risk of overwinding. This danger can be overcome quite easily by marking a few inches of the line with chalk when the fully wound position has been reached.

Keep a supply of assorted weights for adjusting the rate. The weights are added to, or removed from, the pendulum depending on whether the error is a gain or loss. This is explained in detail in Chapter XVIII.

Sometimes work is done in such a position that if a tool was to be dropped, it might fall a long way. In addition to the inconvenience of having to retrieve it, there is the danger that somebody underneath might be hit. Hence it is advisable to tie one end of a length of string to the tool, and the other end of the string to a convenient fitting.

Some of the old clocks are constructed from wrought iron in the form of a box type open frame. Part of the frame-work consists of vertical members which are held in place by nuts and bolts. These vertical members are drilled to take the pivots, and by removing the nuts and bolts the bars can be removed and the wheels will drop out.

The very early clocks had their pivot bushes riveted to the vertical bars, but in later years it became more common to secure the bushes by screws. To rebush a pivot hole, it is necessary to have the new bush specially turned.

If the leading off wires show any signs of wear, new ones should be made and fitted at the first opportunity. Inspect the hammer springs. If they are worn or badly adjusted they can cause the hammer to rest too close to the bell. In this position, it is possible for the hammer to strike the bell twice which may crack it.

It is extremely undesirable that the bells should ring when resetting or checking the strike mechanism, and so by pressing down on the hammer tails one can feel the mechanism in operation.

To take a clock apart, first remove the weights and the pendulum. Then remove the connecting rods from the bevel wheel (if it has more than one dial) by knocking out the taper pins. The leading off wires from hammer tails to hammers can then be removed, leaving only the going and striking parts to dismantle from the frame.

XVII

400 Day Clocks

As the name implies, these clocks will function for 400 days, or approximately one year, at one full winding of the mainspring. They are designed to be decorative as well as timepieces. Almost all the parts are made of brass, and they are polished and lacquered. The movement is mounted on a bridge, which is secured to a brass base that carries four adjustable feet. At the back of the movement is a rotary pendulum made up of two, three or four spherical weights. The pendulum is suspended by a length of spring rather like that which is used for large hairsprings. The pendulum rotates about three quarters of a complete revolution and then changes direction. Clamped to the top of the pendulum spring, and just below the point of suspension, is a fork that engages a pin in the pin pallet arbor.

To prevent damage to the pendulum during transit an arm that protrudes from beneath the base is moved sideways. A concave conical piece is raised through a hole in the base until it supports the bottom of the pendulum, and arrests further movement.

Each spherical weight is mounted on an arm which is pivoted at the top. If the weights are moved outward, the movement rate will be slowed down, and if the weights are brought in closer to the center, the rate will be increased.

There are various methods of adjustment, one of which is illustrated in Fig. 132. By turning the knurled disc in a clockwise direction from the top, the carrier is moved downward, causing the weights to move outward and the movement to slow down.

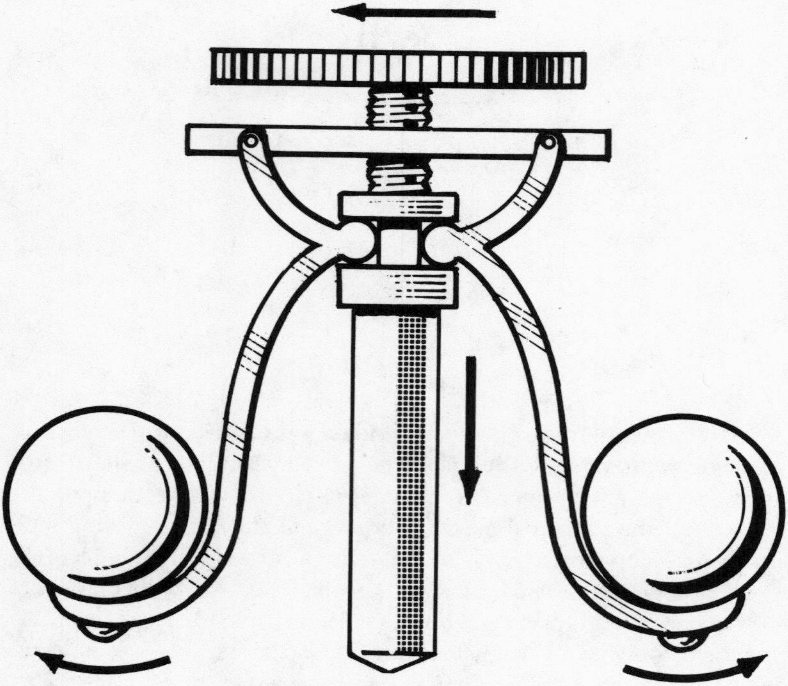

Fig. 132 Regulating 400 day clock

A glass cover fits over the movement and the dial, and serves the dual purpose of maintaining the mechanism free from dust and enhancing the appearance of the clock.

As timekeepers, these clocks can be troublesome if the escapement is out of beat and if the pendulum is out of alignment.

To set the escapement in beat, move the pin pallet slowly until an escape tooth drops, and make a note of the position of the pallet arbor pin that engages the fork. Then slowly move the pin pallet in the opposite direction until the next escape tooth drops. Again make a note of the position of the pallet arbor pin. When the pendulum is in natural rest, the correct position for the fork is midway between these two points.

Release the lock screw above the top of the suspension block, and turn the pendulum spring assembly until the fork is in the midway position, as indicated by the pin pallet arbor pin. Tighten the locking screw and set the clock in motion. If the escapement appears to be dry, put just a touch of clock oil on the tip of each escape tooth and one touch of oil in the pendulum spring fork. These clocks are sensitive to any change in friction in the balance assembly and the escapement.

Adjust the feet to bring the base of the clock into the horizontal position. The pendulum will probably oscillate with a wobble, but this should disappear after a few minutes.

It is necessary to stand the clock on something rigid if it is to function without influence from the shake or vibration that is transmitted through the base.

Cleaning can be carried out by any of the conventional methods, but if brushing is needed then it would be advisable to use a soft brush, and even then brush with the grain. A stiff brush can leave marks in the lacquer and if these marks are across the grain it results in a permanent blemish. The only means of restoration is to clean the affected part down to the metal, regrain and re-lacquer.

XVIII

Pendulums

MUCH OF THE THEORY OF PENDULUMS IS FOR STUDY BY MATHEMATICIANS, and only little of this is needed by the clock repairer. The man at the bench is more concerned with correcting faulty actions and adjusting for timing. Sometimes he has to calculate the length of a pendulum in case the original is lost. It is unfortunate, but nevertheless true, that there is a tendency for pendulums to be taken for granted and not given the attention they deserve, and so a little theoretical knowledge may lead to a better understanding.

If we suspend a small piece of lead about the size of an orange pip from a length of fine nylon thread about 39 in. long, and then attach the other end to something rigid, we would find that if we allowed the weight to swing the time taken from one end of the swing to the other would be approximately one second.

The point at which the thread is suspended is known as the center of suspension, and the center of gravity of the lead weight is known as the center of oscillation. Mathematicians have calculated that to make this simple pendulum swing from one side to the other in exactly one second, the distance between the center of suspension and the center of oscillation has to be 39.14 in. This is known as a one second pendulum.

By moving the weight above or below the original position of the center of oscillation, we vary the time taken to complete a swing.

If we raise the weight, the rate of vibration is increased and the time taken to complete a swing becomes less than one second. If we lower the weight, the reverse effect takes place. This is the near theoretical pendulum upon which mathematical calculations for more practical applications are based.

The formula for calculating the length of any pendulum is:—

$$1 = \frac{\text{secs. pendulum length} \times (\text{secs. pendulum vib./min.})^2}{(\text{replacement pendulum vib./min.})^2}$$

Before the length of the new pendulum can be calculated, it is necessary to know how many vibrations it will make in one minute. This is obtained by calculating the count of the gear train as explained in Chapter V.

Example 1

If we have a clock with a center wheel of 84, a third wheel of 70, and an escape wheel of 30, with third wheel and escape wheel pinions of 7, the count will be:

$$\frac{84 \times 70 \times (30 \times 2)}{7 \times 7} = 7200 \text{ vibrations per hour}$$

which equals $\frac{7200}{60} = 120$ vibrations per minute or

2 vibrations per second. This is known as a half seconds pendulum.

We know that the length of a seconds pendulum is 39.14 in. and so we can now substitute figures into our formula:

$$1 = \frac{39.14 \times 60 \times 60}{120 \times 120} = \frac{39.14}{4} = 9.78 \text{ in.}$$

Example 2

Center wheel 80, third wheel 72, escape wheel 30, and third wheel and escape wheel pinions of 8, the count will be:

$$\frac{80 \times 72 \times (30 \times 2)}{8 \times 8} = 5400 \text{ vibrations per hour}$$

which equals $\frac{5400}{60} = 90$ vibrations per minute.

The formula can now be written:

$$l = \frac{39.14 \times 60 \times 60}{90 \times 90}$$

$$= \frac{39.14 \times 2 \times 2}{3 \times 3} = \frac{156.56}{9} = 17.39 \text{ in.}$$

It must be remembered that these lengths are theoretical distances between the center of suspension and the center of oscillation. The actual length of the pendulum will be approximately an inch longer, depending on the weight and shape of the bob and the weight of the rod.

An inexpensive but efficient seconds pendulum for a grandfather clock can be made from a length of wooden rod and a lead weight, Fig. 133. The finished pendulum must measure 45 in. from the point of swing to the bottom of the lead bob, and the wooden rod with its brass fittings has to be made with this dimension in mind.

A good rating can be obtained by using straight grained, well-seasoned boards of fir or pine for the rod. Select a piece that is free from knots and shakes, and shape a rod ½ in. diameter.

Obtain a length of brass tubing that will fit tightly over the rod and cut it into three lengths, measuring approximately 2 in., 1 in. and 1½ in. Into the 2 in. length, insert a tight fitting piece of brass rod 1 in. long to a depth of ½ in., and soft solder the two together. This is described in Chapter II. File the end to shape, and cut a thin slot to receive the suspension spring. Push the fitting firmly onto the end of the wooden rod, and secure it by drilling and pinning with thin brass or steel rod. The saw cut must be straight and square with the pendulum rod, and ideally the width should be just sufficient to admit the thickness of the suspension spring. A circular saw set up in a lathe is probably the most accurate way of doing this.

The 1 in. length of tube is pushed onto the wooden rod to a position opposite the crutch. The piece of tube is secured by pinning, and then a slot is formed by drilling and filing from one side of the rod to the other. The width and length of the slot should be sufficient to provide the impulse post of the crutch with complete freedom of movement, without having too much play. It is essential that the crutch slot be parallel to the suspension spring slot. If it

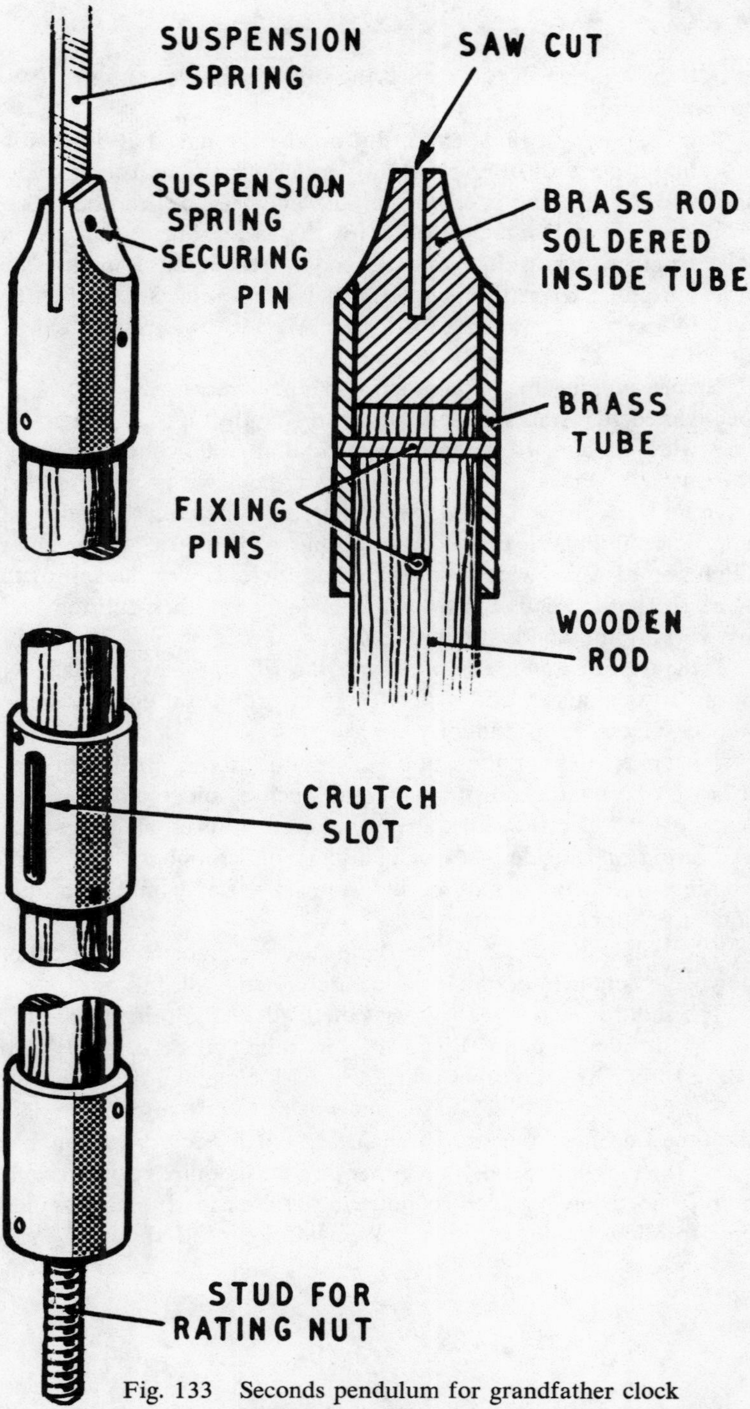

Fig. 133 Seconds pendulum for grandfather clock

is not, the crutch will rub against the sides of the slot and the movement will stop.

The remaining 1 in. length of brass tube is plugged at one end by inserting a piece of brass rod to a depth of ½ in. and then soldering. The blanked end is center-marked and then drilled and tapped to accept a $\frac{3}{16}$ in. screw thread for the rating nut at the bottom. The length of protruding thread needs to be about 1 in. to allow for shortening when the pendulum has been adjusted. The easiest way to insert a screw thread is to put in a 1½ in. screw to a depth of ½ in., and then cut off the head.

Before pinning to the wooden rod, the three brass ferrules can be cleaned by using very smooth emery cloth. Open the jaws of a vise wide enough to support the pendulum rod without having it fall through. Place a piece of rag over the jaws to prevent them from marking the wood and, with an assistant holding the rod firmly and horizontally, clean the brass fittings with a strip of the emery cloth about ½ in. wide. Wrap the emery cloth over the top of the work so that the ends hang down each side, and then rub the brass by working the hands up and down.

If the wood is sanded smooth, and varnished, the appearance will be greatly enhanced and will offer good protection against the influence of changing humidity.

To make the pendulum bob will require about 20 lbs. of lead. This can be melted down from scrap such as old plumbing pipes, lead-covered electric cable, strip or sheet lead used as damp-proof protection on outside walls of buildings, and roofing sheets. If old buildings are being demolished in your area, this would be a likely source of supply.

The bob is to be 12 in. long and 2¼ in. diameter with a hole down the center into which the pendulum rod will fit.

The mold in which we will cast the lead needs to be a tube of a 2¼ in. inside diameter, Fig. 134. The tube can be of thick cardboard rather like the tubes that are used for mailing drawings, etc.

Cut two discs of thick card that will fit tightly into the ends of the tube and in the center of each disc cut a circle big enough to admit the pendulum rod. The upper disc will require two additional holes, one through which to pour the molten lead, and a smaller hole to allow the air to escape. We will also need about 15 in. of wooden rod the same diameter as the pendulum rod for the core of the mold. Fix the two discs in position and pass the length of wooden rod down through the center of the tube until it protrudes about 1 in. at the bottom.

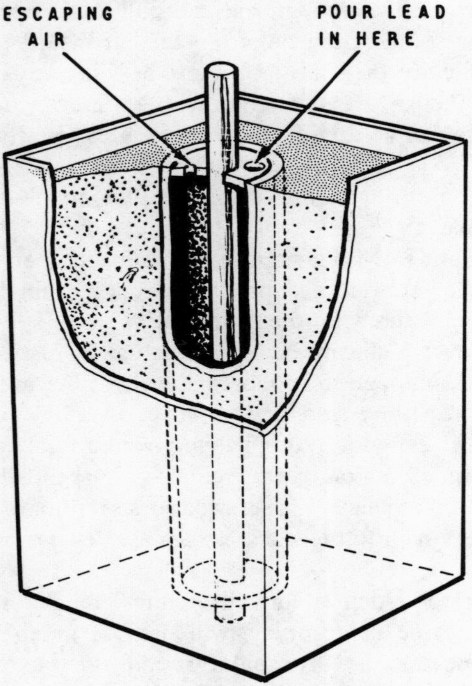

Fig. 134 Casting a lead bob

We will need a box a little over 12 in. deep. Place the mold vertically in the center, and tightly pack it with damp sand underneath and all around until it is level with the top of the mold.

Lead has a low melting point, so there is no problem in melting it down. The pieces of lead are placed in an old saucepan and heated over a gas ring or electric hot plate. WARNING: UNDER NO CIRCUMSTANCES SHOULD THIS BE CARRIED OUT IN A SAUCEPAN THAT IS USED FOR THE PREPARATION OF FOOD BECAUSE OF THE RISK OF LEAD POISONING.

Carry the molten lead into the open air and pour it very carefully into the mold until it is full. The lead will set in a few minutes but will remain hot for sometime afterwards. There is no reason why the contents of the box should not be tipped out onto the ground to speed up the cooling process.

Twist and wriggle the wooden rod and pull it out from the lead. The lead casting should push out from the cardboard tube, but if there is any difficulty then the tube must be cut away. Take care not to disfigure the surface of the bob with knife cuts. Incidentally, the same process can be used for making weights for weight-driven clocks.

The suspension spring is fitted into the slot and is drilled and pinned in position.

The bob is trimmed with a smooth file, and slid over the pendulum rod until the screw thread protrudes. Fit a plain flat washer over the thread and screw on the rating nut.

The cock from which any heavy pendulum is suspended must not only be strong enough to support the weight, but must be sufficiently rigid to resist any tendency to move. Even the best pendulums will fail to give a good rating if hung from a weak suspension.

The pendulum of a grandfather clock is suspended from the movement, and the movement is screwed to a seat board supported by wooden cheeks held to the sides of the case. This is not the best arrangement.

A precision clock, such as an astronomical regulator, designed to give a high degree of accuracy, will provide independent suspension for the pendulum. It is usual to support the movement on a cast iron bracket screwed to the back of the case, and the pendulum suspension cock is frequently screwed to this bracket. Such an arrangement allows for the removal of the movement so the pendulum can remain in the case. This is a great advantage when working on a clock fitted with a pendulum weighing as much as 30 lbs.

With any pendulum, the contact between the crutch and the pendulum rod must be kept to a minimum to avoid unnecessary friction. The center of suspension of the pendulum has to be directly opposite and in line with the pallet staff pivot. The pendulum must swing along a path that is at right angles to the pallet staff. There must be no tendency to turn or wobble during the swing. If there is, the trouble can frequently be traced to bad alignment at the suspension, or a damaged or weak suspension spring. If the suspen-pension spring is bent or creased it cannot be properly straightened, and the only satisfactory remedy is a new spring.

To set the pendulum in beat, wind the spring a few turns if it is a spring wound clock, or wind up the weights. Slowly move the pendulum to one side until an escape wheel tooth drops onto the pallet. Make a note of the distance the pendulum has swung from the vertical position, and then move the pendulum in the opposite direction until another escape wheel tooth drops. If the distance traveled by the pendulum is the same on each side of the vertical, then the pendulum is in beat. If the pendulum is not in beat it will have swung through an angle greater on one side of the vertical position than the other. All that is necessary to correct this is to bend the crutch in the direction of the greater angle.

A rise in temperature will cause a pendulum to expand, which will increase its length. The center of oscillation will be taken further from the center of suspension, which will reduce the rate of vibration and cause the clock to lose time. A drop in temperature will have the reverse effect. It is for this reason that temperature compensating pendulums are necessary in a precision-built clock. There are many different types of compensation pendulums, some of which have seen little or no change in design since they were invented. Some of the more common ones are described in the following paragraphs.

THE ELLICOTT PENDULUM: (Fig. 135) The pendulum rod is made of steel. It passes right through the bob and has a supporting nut on the end. Secured to the top of the pendulum rod is a bridge into which two brass rods are screwed. The lower ends of these brass rods are in contact with two rocking arms that pivot in the bob. When the pendulum is subjected to a rise in temperature the steel rod expands and the bob is lowered.

At the same time that this is happening, the brass rods are expanding, and in so doing they press down on the inner ends of the rocking arms. This causes the outer ends to lift and raise the bob on the lifting pins.

Because the coefficient of expansion (expansion rate) of brass is greater than that of steel, the brass rods gain on the steel rod and

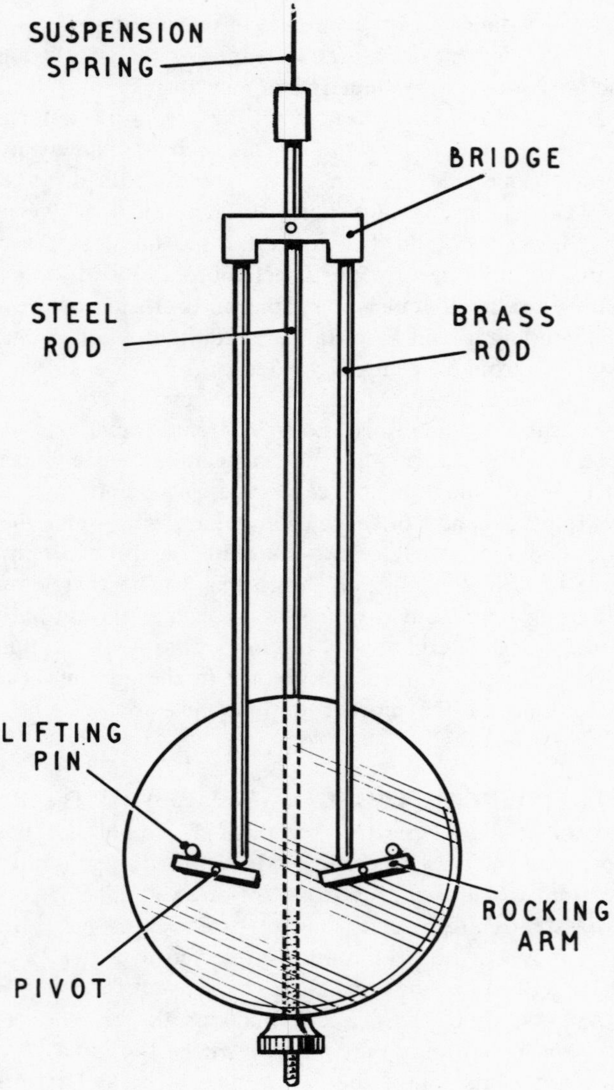

Fig. 135 The Ellicott pendulum

are able to raise the bob enough to maintain the original position of the center of oscillation.

THE ELLIOTT PENDULUM BOB: (Fig. 136) As will be seen from the illustration, most of the metal of the bob is above the adjusting nut. When the pendulum rod expands downward, the bob expands a compensating amount upward and the center of oscillation remains unchanged. Because of the different expansion rates, the position of the nut will vary according to the metal used.

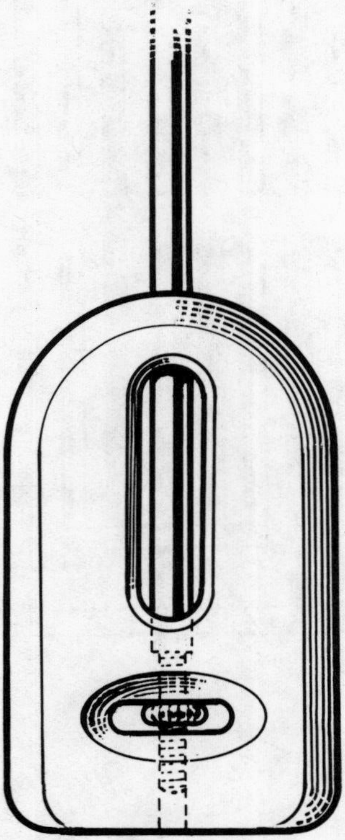

Fig. 136 The Elliott pendulum bob

THE GRID IRON PENDULUM: (Fig. 137) The pendulum consists of an assembly of steel and brass rods, alternately placed so that the expansion effects cancel out one another.

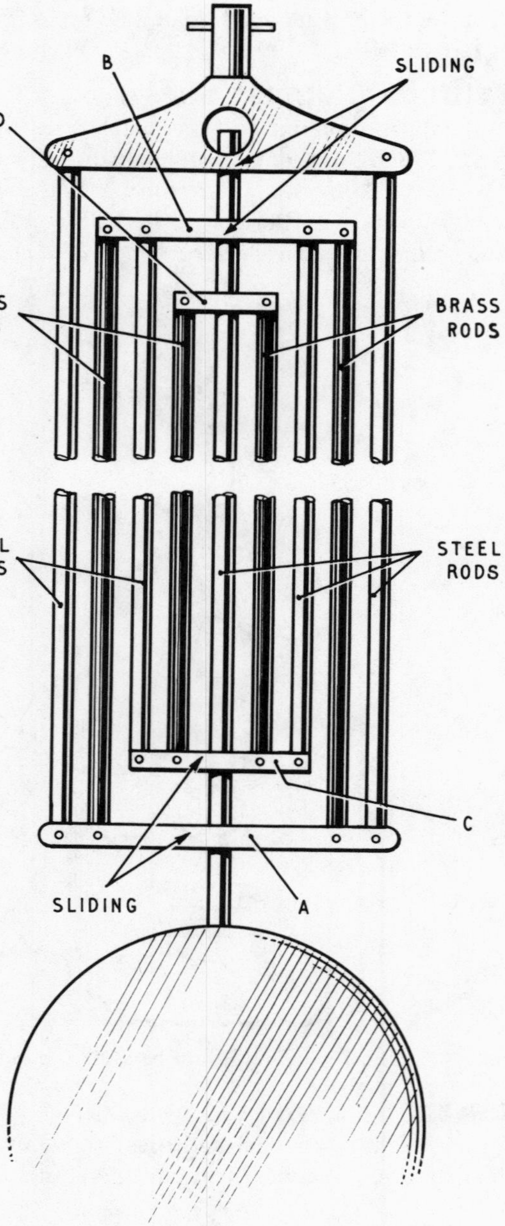

Fig. 137 The grid iron pendulum

A bridge piece forms the top of the pendulum, and suspended from each end is a steel rod. The lower ends of these rods are secured to a horizontal bar A that supports two brass rods. A rise in temperature causes the steel rods to expand downward, taking with them the horizontal bar A and the two brass rods. But the brass rods expand upward, taking with them horizontal bar B from which two steel rods are suspended. This second pair of steel rods expands downward, lowering horizontal bar C to which is attached a second pair of brass rods. These brass rods expand upward, raising horizontal bar D and, because the bar is fixed to the center pendulum rod, the pendulum bob is also raised. The effect of these different expansions is to stabilize the bob, and maintain the position of the center of oscillation.

The lower end of the pendulum rod is threaded to take a rating nut, and it is also machined to a fine point which, when the pendulum swings, acts as a pointer in front of a graduated scale secured to the back of the clock case.

The illustration shows the pendulum in a broken diagrammatic form for clarity. In reality, the pendulum is quite long and the rods are very close together to form a neat and compact assembly. The pendulums are heavy, usually between 20 lbs. and 30 lbs., and need to be handled with care. Lift them from their suspension and lay them on the floor with the point towards the wall, where they are out of harms way.

Some clocks are fitted with imitation grid iron pendulums; they consist of an assembly of rods which are not compensating.

THE INVAR PENDULUM: The word invar is the name given to a nickel-steel alloy that has a very low coefficient of expansion. This characteristic makes it very suitable for pendulum rods, particularly when they are fitted with compensating bobs as shown in Fig. 138.

This type of bob is made of cast iron or mild steel and is cylindrical in shape. The center hole in the upper half is just large enough to admit the pendulum rod, but in the lower half the hole is enlarged to receive a length of brass tube that closely fits around the pendulum rod, but is still free to move.

The lower end of the brass tube rests on the rating nut, and the

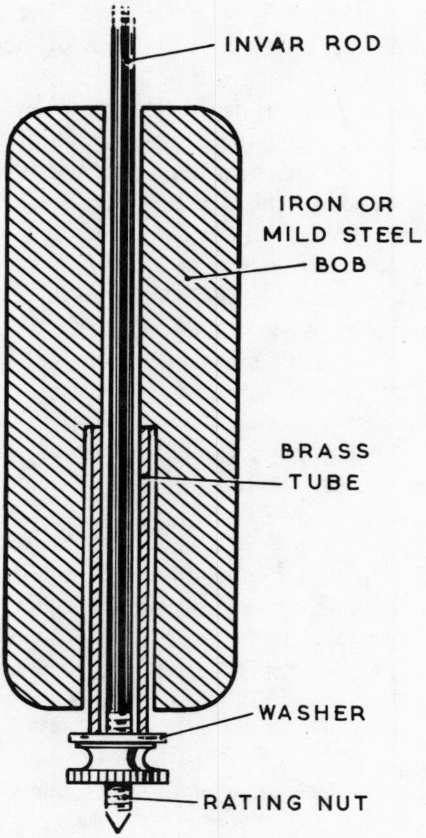

Fig. 138 The invar pendulum

upper end of the brass tube supports the bob on the shoulder of the recess, midway along the hole. The bob expands or contracts an equal amount above and below the point of support because it is supported in the center, and thus maintains the center of oscillation.

MERCURIAL PENDULUM: (Fig. 139) This is a pendulum capable of accuracy to within one second per month. It consists of a steel rod supporting an iron or glass vessel containing mercury (quicksilver). When the steel rod expands downward due to a rise in temperature, the mercury expands upward and compensates.

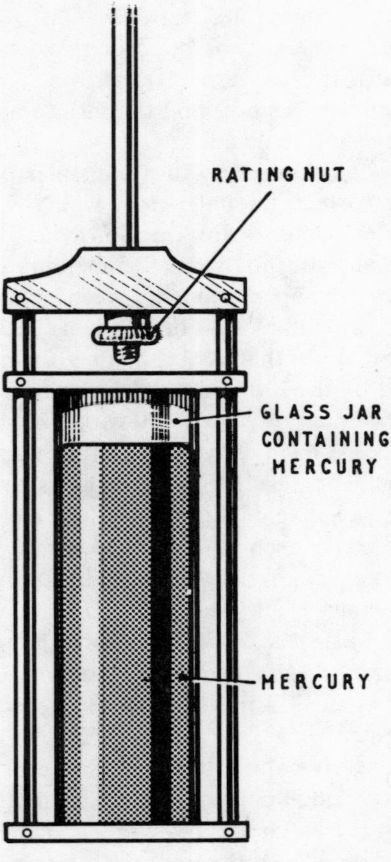

Fig. 139 Mercurial pendulum

Glass is not a good conductor of heat, and when it is used to hold the mercury the steel rod expands downward before the mercury is aware of the temperature change. This is not so with iron because of its sensitivity to changes in temperature. Anyway, iron is the only metal that can be used because any other metal would be attacked by mercury.

If the pendulum bob can be seen when installed, a glass container is undoubtedly preferred because of its appearance. With astronomical regulators, where first consideration is to the provision of accuracy, and appearance is of secondary importance, it is usual to fit a bob with an iron vessel. These iron vessels are accurately machined to ensure balance, and the insides are usually enamelled to seal the pores of the metal.

The amount of mercury in a container will vary from clock to clock. The correct amount is that which will expand upward just enough to compensate for the downward expansion of the rod. During initial tests, the quantity of mercury in the vessel is adjusted until accurate compensation has been achieved. This is indicated by keeping the rate of gain or loss constant throughout changes in temperature. When this has been done, the center of oscillation can be adjusted by the rating nut to give accurate timekeeping.

Some pendulum rods are fitted with a friction tight brass disc just above the bob, on which weights can be placed while the pendulum is in motion. These weights are usually as small as those that might be used for weighing mailing letters. If the clock or regulator is losing a few seconds each month, and a small weight is placed on the platform it may be that at the end of the following month the movement is gaining a few seconds. The first weight is removed by tweezers and is replaced by a lesser weight and so on until accurate timekeeping has been achieved. The famous clock Big Ben at Westminster, London, is adjusted in the same way, using copper coins as the weights.

One frequently comes across French short pendulum clocks using mercurial pendulums for appearance only. Usually the mercury is contained in two glass tubes with end caps as shown in Fig. 140. With these bobs, the rod passes right through and the rating nut is at the bottom.

Beware of imitations. Some short pendulum clocks are fitted with what appears to be twin glass tubes containing mercury, but in fact contain a short length of steel tube with a convex disc soldered to the upper end and chrome plated. It is difficult to distinguish from mercury by a casual glance.

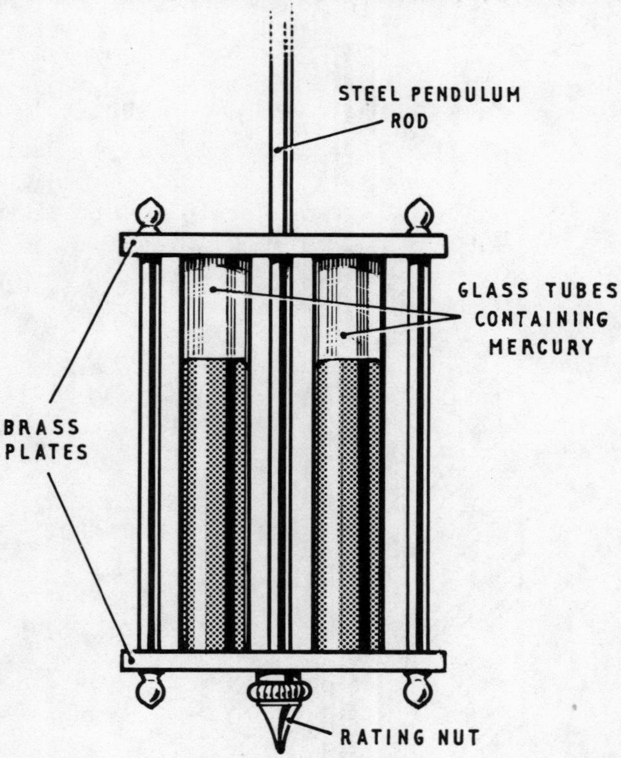

Fig. 140 Mercurial pendulum

ZINC AND STEEL PENDULUM: (Fig. 141) This arrangement produces very accurate results. Many famous public building clocks throughout the world are fitted with this type of pendulum.

The pendulum rod is iron and its lower end is threaded to receive the rating nut. Resting on the rating nut is a recessed collar with a hole a little larger in diameter than the rod. A zinc tube which can slide over the rod sits in the recess of the collar.

Surrounding the zinc tube is a steel or iron tube. Secured to the top of the steel tube is an internal collar that rests on the top of the zinc tube, and secured to the bottom of the steel tube is an external collar that supports the lead bob at its halfway point.

The coefficient of expansion of iron and steel is almost the same, so it matters little which is used: it depends more on availability. The coefficients of expansion between iron and zinc are, however,

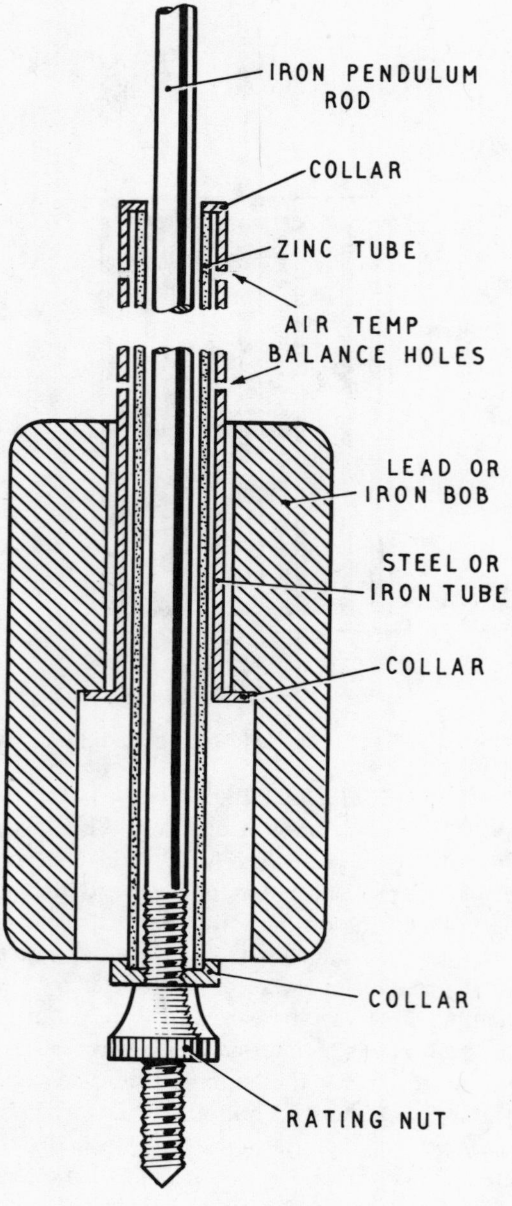

Fig. 141 Zinc and steel pendulum

very different. Zinc and lead both expand a little more than twice that of iron and steel under identical conditions. Within the temperature range of 0°*C* to 100°*C*, the coefficients for iron and zinc are 12 and 28 respectively.

The iron pendulum rod expands downward taking with it the zinc tube, but at the same time, the zinc tube expands upward and, with its higher coefficient of expansion, it over-corrects. This is cancelled out by the downward expansion of the outer iron or steel tube, which keeps the bob stationary. The bob is supported in the center, and so any change in dimension is divided equally above and below the center of oscillation.

Holes are drilled in the outer tube to allow any change in air temperature to pass to the zinc tube.

This is not a difficult pendulum to make, and many clock material dealers will be able to supply the lengths of rod and tubing needed. Dimensions will be influenced by availability of material, but suggestions for a seconds pendulum are:

lead bob—8 in. long × 2½ in. diameter
zinc tube—24 in. long
iron tube—20 in. long
or
iron bob—8 in. long × 2½ in. diameter
zinc tube—28 in. long
iron tube—24 in. long.

Both of these use an iron pendulum rod 46 in. long × ¼ in. diameter.

XIX

Floating Balance

IT IS THE BALANCE OR PENDULUM OF A CLOCK MOVEMENT THAT has the greatest influence over maintaining accuracy of a timepiece. They must be allowed to function with as little frictional resistance as possible, and in good poise. To achieve this it is essential that their pivots are designed so as to offer the minimum bearing surface, and it is here that most of the troubles lie. To reduce frictional resistance the pivots are machined as small as possible, without making them too weak for normal function. Because the weight of the balance or pendulum remains constant, it follows that with a decrease in bearing surface there will be a proportionate increase in the rate of wear. This in turn makes the pivots more prone to breakage in the event of a shock.

Fig. 142 illustrates the floating balance which was designed to overcome these problems. The balance staff has been replaced by a tube, and into each end is pressed a jewel hole. Passing through the tube is a length of spring steel wire or piano wire, which is pulled taut and anchored on the upper and lower arms of the support bracket.

The balance spring is no longer used. In its place is a helical spring which is secured at its lower end by a collet on the tube, and at its upper end by a collet which is part of the bracket assembly and is therefore stationary.

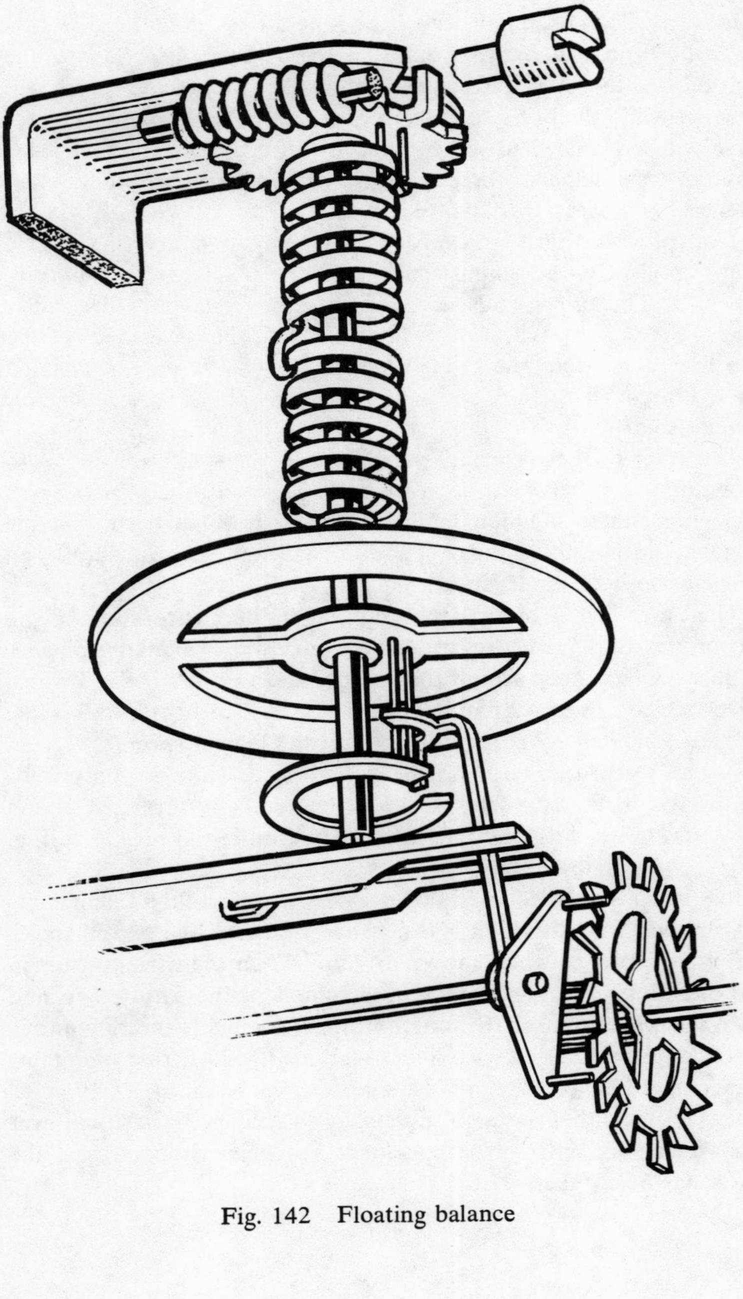

Fig. 142 Floating balance

Midway along its length the spring is bent back upon itself so that the coils of one half are wound in the reverse direction to the coils of the other half. During oscillation, one half of the spring will contract while the other half will expand an equal amount. The reverse will happen when the balance wheel changes direction. This maintains the balance wheel in a horizontal attitude and there is no tendency to ride up or down the wire.

The roller has been replaced by a C shaped safety ring, and a pair of polished steel impulse pins take the place of the ruby pin, Fig. 143. The upper ends of the pins are pressed into the under face of the balance wheel near its center, and the lower ends of the pins are driven into the safety ring, holding it firmly but isolated from the rest of the balance assembly. Two pins are necessary to prevent the ring from rotating.

The lever of the pin pallet escapement is bent at right angles to allow the notch between the horns to engage with the impulse pins while the pallets function in a vertical plane. One horn is made longer than the other, and this additional length is bent parallel to the main body of the lever.

The lower arm of the balance support bracket is forked and the two prongs thus formed carry out the function of banking pins.

Beneath the upper arm of the balance support bracket is a worm wheel which carries a pair of polished steel curb or index pins. Between this passes the topmost coil of the helical spring.

In mesh with the worm wheel is a worm cut integral on an adjusting rod with a screwdriver slot at one end. By turning this rod the worm wheel will carry the curb pins around the helical spring, and alter the timing rate.

It is the physical isolation of the safety ring from the balance assembly that allows the long horn to pass through the gap in the C ring as the balance wheel moves around. When the wheel moves in one direction, the long horn is imprisoned in the safety ring and when the wheel returns the long horn passes outside the ring and is freed. It is this safety device, illustrated in Fig. 144, that maintains engagement between the impulse pins and the lever.

It is unlikely that the wire supporting the balance wheel will ever need replacing, unless the movement has been dropped and the shock has been taken by the balance.

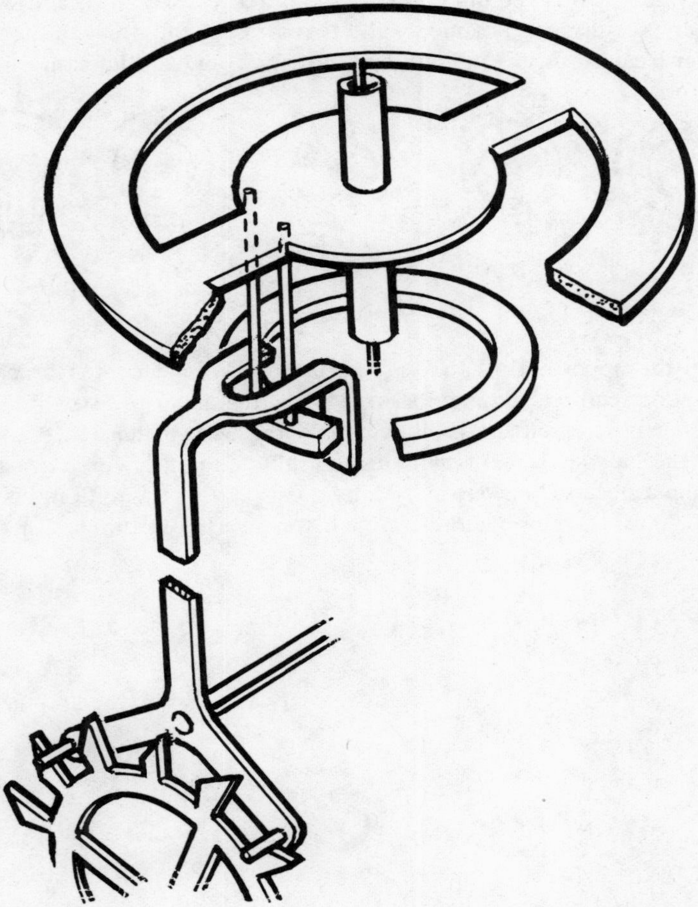

Fig. 143 Floating balance index

Release the wire from the bracket by levering up the tags holding the wire ends. Cut the wire close to where it enters the upper arm of the bracket, and pull the two pieces of wire from the movement.

Bend the replacement wire to ninety degrees and pass it upwards through the tube. Wrap the wire around the lower fixing tag, and bend the tag downward to secure the wire. Make sure the new wire is laying flush on the underside of the bracket lower arm, and with finger pressure gently squeeze the two arms of the bracket slightly toward each other. Maintain this pressure, and at the same time

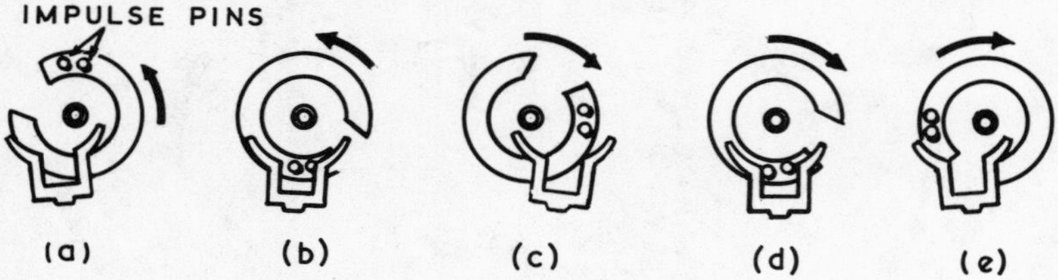

Fig. 144 Floating balance cycle

bend the upper end of the wire so that it lays flat on the bracket arm, and secure it to the upper tag. When the finger pressure on the bracket is released the arms will hold the wire in tension. Check that the balance wheel is floating centrally along the wire. Any adjustment necessary is carried out by gently easing the bottom collet up or down the tube with a screwdriver inserted in the gap of the collet.

XX

Battery Clocks

A BATTERY-OPERATED CLOCK DOES NOT HAVE TO BE WOUND, THERE are no wires to connect it to a main electricity supply, it is portable, and no precautions need be taken against receiving an electric shock during servicing or when testing. The movements are small and compact and are easy to service. A 1.5 volt leakproof battery powers the movement and should last for at least twenty-four months.

It is not necessary for the watch or clock repairer to have a knowledge of electronics before attempting to service a battery-operated clock movement. The mechanical section is simple in design and will present few, if any, problems. There is little that can go wrong with the electronics other than battery replacement, and possibly renewal of balance parts or the complete balance assembly. Manufacturers usually design their movements so that the electronic section can be separated from the mechanical section.

In a conventional spring wound movement the motive power is stored by the mainspring, and is released under control through a train of wheels to terminate at the balance. Power to rotate the hands is taken from the wheel train.

In battery-operated clock movements this is not necessarily so. Some manufacturers reverse the direction of drive. The balance wheel is made to be the source of mechanical energy, and its oscillations drive the escapement, which in turn drives the train. Such a movement is said to have a driving escapement.

Other manufacturers are more conventional in their designs. Some employ a spring to operate a train and balance in the usual way, and arrange for the spring to be rewound by electrical means. To be able to do this with a 1.5 volt battery necessitates the use of a lightweight spring. Such a spring can store only a very limited amount of energy, and after about two minutes it is in need of rewinding.

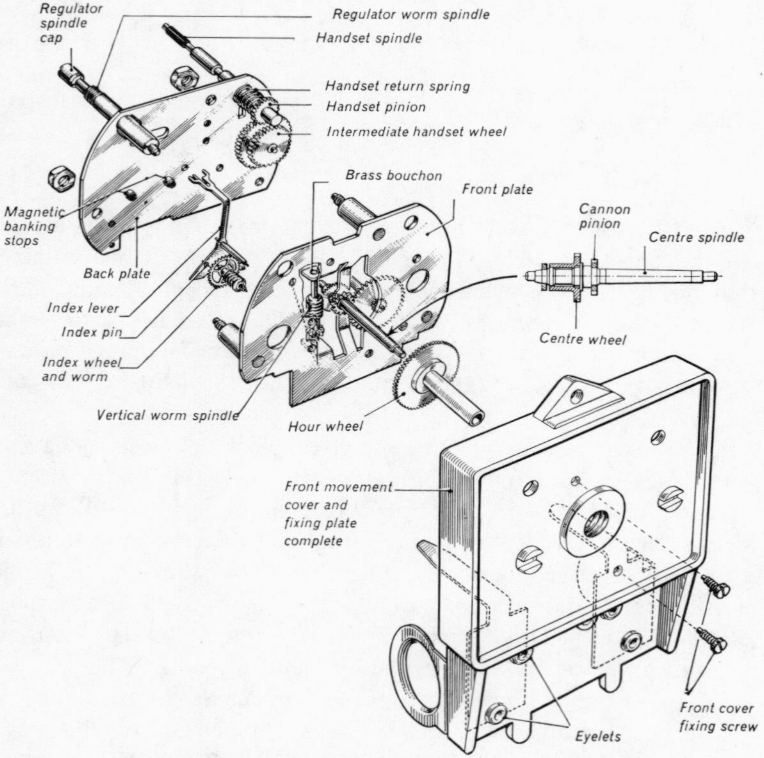

Fig. 145 Clock mechanism of Smiths transistorized movement

Fig. 145 shows an exploded view of the clock mechanism of a Smiths transistorized battery clock movement, and Fig. 146 illustrates the electronic section. This movement has a driving escapement, and the energy from the battery is released through an electronic circuit to cause the balance wheel to oscillate. These oscillations are used to drive a geared index and a wheel train, and power is taken from this to move the hands.

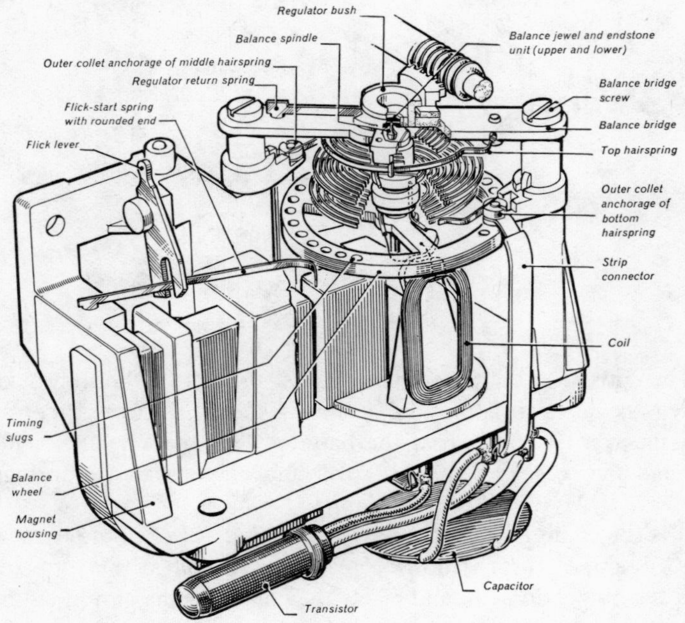

Fig. 146 Electronic section of Smiths transistorized movement

Mounted on the balance staff are three balance springs, and secured at right angles to the rim of the balance wheel is a coil.

The electronic circuit is illustrated at Fig. 147 and it will be seen that the three balance springs are electrically connected. A transistor, a capacitor, and a battery complete the circuit.

The coils consists of a drive coil and a trigger coil, which are wound one on top of the other. The coils are connected at their inner ends, at which point they make electrical contact with the center balance spring. The outer ends of the coils are connected to the top and bottom balance springs, respectively.

When the balance wheel oscillates, it carries the coil through the magnetic field of a permanent magnet, setting up an induced current in the trigger coil. This current alternates from positive to negative as each leg of the coil passes through the field.

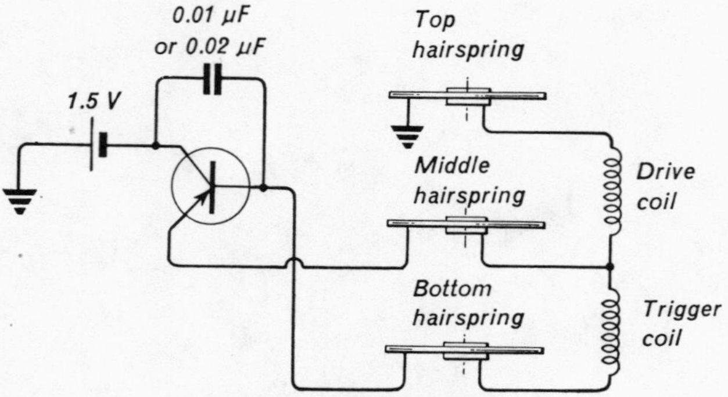

Fig. 147 Electronic circuit

The current in the trigger coil passes through the bottom balance spring to the transistor, and switches the transistor to permit a larger current to flow from the battery. This battery current flows through the transistor, and the top and center balance springs to the drive coil.

It is the trailing leg of the trigger coil that switches the transistor, and the current that is flowing in the drive coil at the time reacts with the magnetic field, and ejects the coil from the gap in the magnet.

Exactly the same thing happens when the balance wheel returns to complete the second half of its cycle. It is these electrical impulses, generated by the transistor circuit, that sustain oscillation.

The oscillations are counted by an index lever that is bent at right angles, Fig. 148. The upper end is forked in a manner similar to that of a watch lever, and engages with a molded plastic safety roller and impulse pin mounted on the balance wheel staff.

The lower end of the index lever carries two pallet pins that alternately engage with the index wheel and push it around one tooth at a time. If the index wheel should move backwards out of the path of the incoming pallet pin, the wheel is corrected by being pushed forward by the outgoing pallet pin. A friction spring, which bears against the index wheel spindle, holds the index wheel steady.

Mounted on the index wheel spindle is a worm that engages with a plastic worm wheel, and forms part of a molded spindle carrying a worm at its upper end. This upper worm engages a train of plastic gears that follows a more conventional pattern.

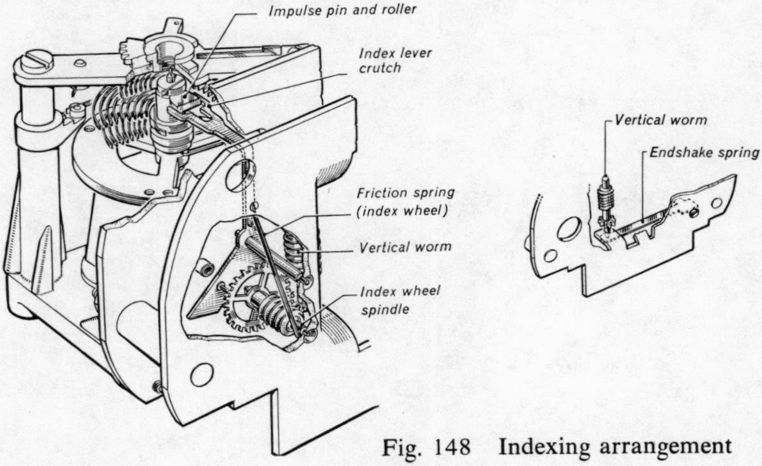

Fig. 148 Indexing arrangement

If the movement fails to function, try a new battery, and check that the hands are not obstructed.

The movement is mounted on a plastic front plate, and is protected by a transparent plastic cover. The combined handset and starting knob can be clearly seen in Fig. 149. By exerting a straight pull the knob can be withdrawn from its serrated spindle. Remove the cover screw, depress the top of the cover, and the cover will lift away leaving the balance clear for inspection.

Unsolder the two battery wires from the battery carrier. Remove the two screws shown in Fig. 150, and then the electronic section can be lifted from the clock mechanism. If the screwdriver touches the permanent magnet, the screwdriver will become magnetized. Once this has happened, the screwdriver should be kept apart from all other steel tools, particularly tweezers. Under no circumstances should it be allowed to touch a watch or clock balance spring, in case the spring is made of a steel that can be magnetized.

Having removed the electronic section, inspect the train of gears for freedom of movement. Each gear should have endshake, sideshake, and backlash with its neighbor. Check that the pallet pins

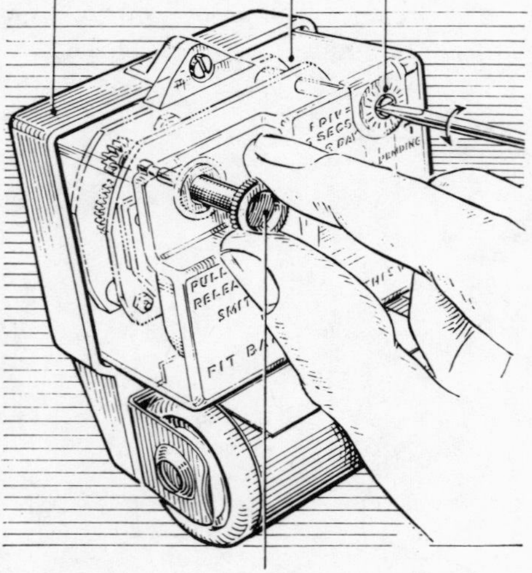

Fig. 149

are not broken, and that the lever is indexing the wheel with each movement.

The clock mechanism is held to the front plate by two screws. Remove the screws and the gear train can be lifted away.

Metal components can be washed in proprietary cleaning fluids, but plastic components should be washed in carbon-tetra-chloride (CCl_4). Normal oiling is carried out except where there is a plastic to metal bearing. In this case, it should be left dry.

Fig. 151 illustrates Smiths standard test equipment, with which the electronic circuit can be quickly checked. If this is not available, and the circuit is suspect, then it is advisable to fit a replacement tagboard.

Unsolder the two insulated metal straps from the left hand and center pins, and the battery wire from the right hand pin, as shown at points A, B and C in Fig. 152. Remove the tagboard screw and lift out the tagboard, complete with capacitor and transistor.

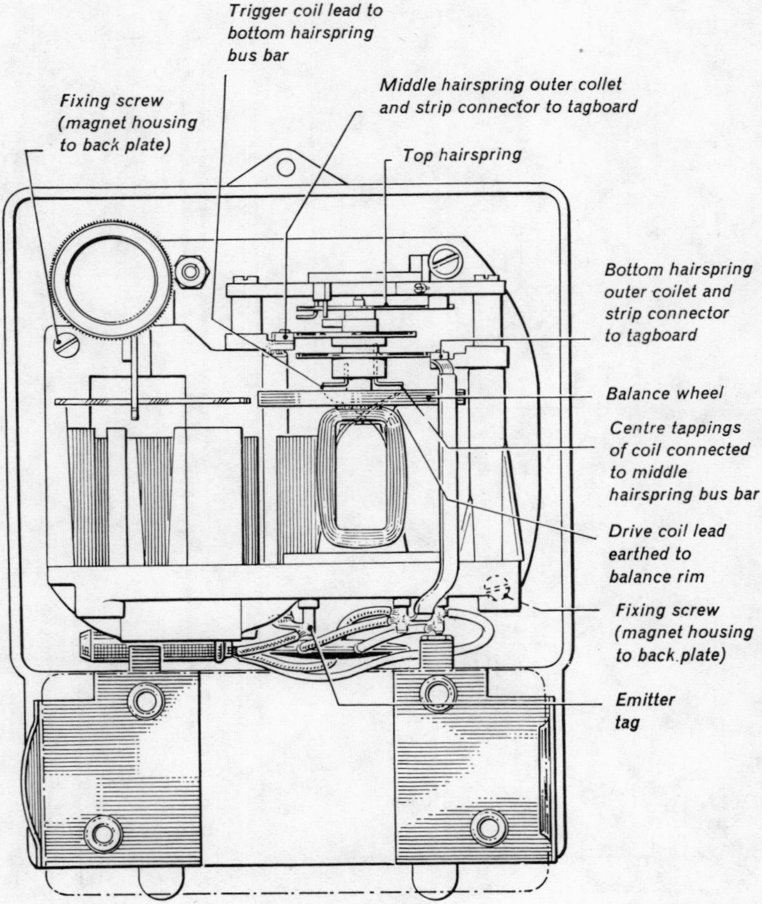

Fig. 150

If the balance staff is broken, or if the coil is suspect, it is recommended that a complete replacement balance be fitted.

Remove the two balance bridge screws, Fig. 146. Release the anchor post for the top balance spring by slackening the side screw, and gently ease the collets of the center and bottom balance springs from their studs. The balance assembly can now be removed, but

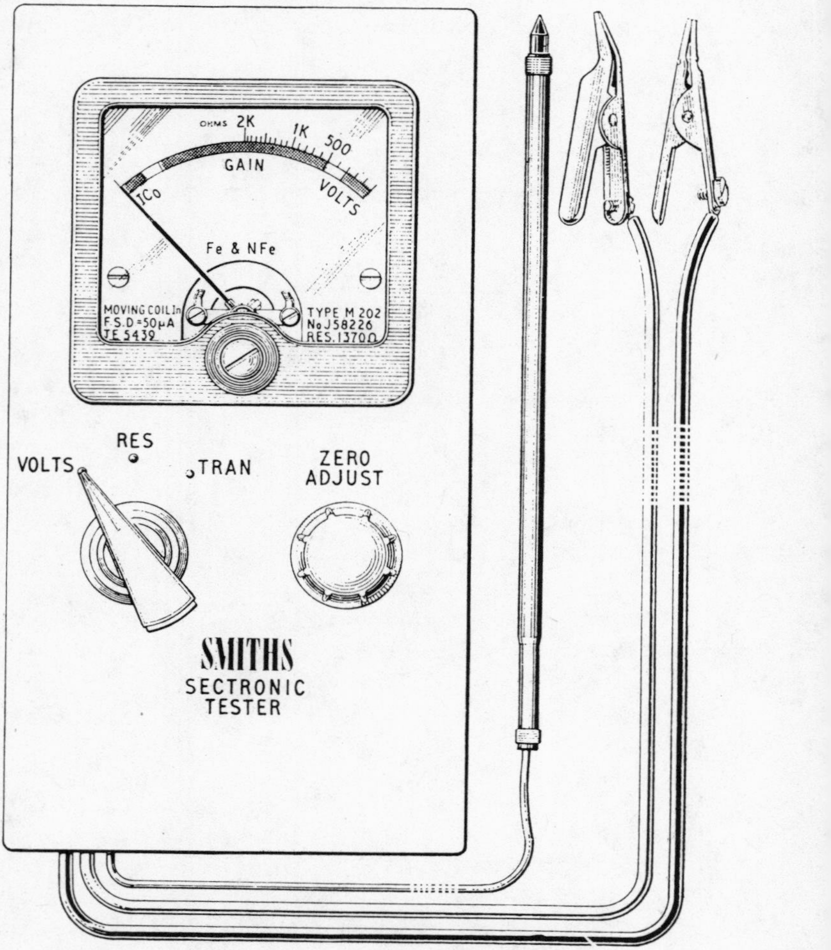

Fig. 151

care must be exercised to avoid damage to the delicate wires of the coil.

Reverse the procedure when fitting the replacement balance, and refer to Fig. 153 for the method of securing the balance springs.

In the Kienzle movement a mainspring is used to give power to the wheel train, but electrical energy is used to wind the mainspring.

The current from a 1.5 volt battery energizes an electro-magnet,

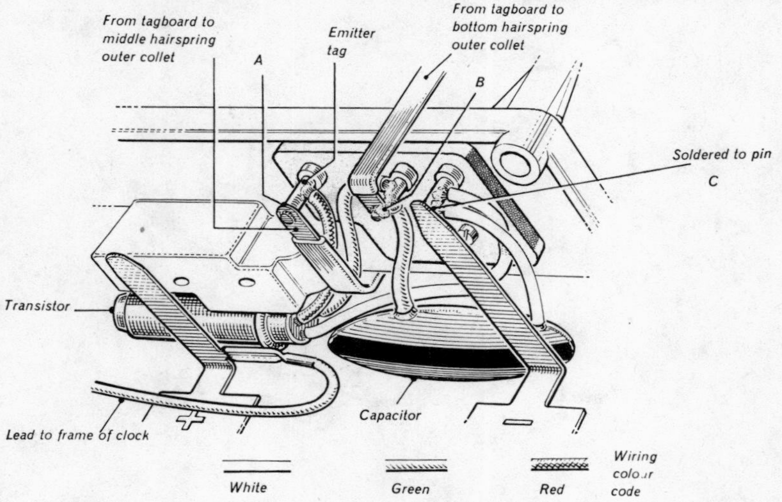

Fig. 152

Fig. 154, which causes an armature hinged beneath the magnet to jump upward. In so doing, it strikes a winding arm, which is part of the mainspring barrel assembly. The impact causes the winding arm to be thrown upward, which turns the barrel and winds the mainspring.

Two electrical contact points, one on the armature and the other on the winding arm, are now separated, the electrical circuit is broken and the electro-magnet becomes de-energized. In this condition, the armature drops to its former position.

Pivoted on the end of the winding arm is a spring-loaded pawl, that engages the teeth of a ratchet wheel that rotates on the same axis as the barrel.

When the winding arm is given an impulse the pawl rides over the ratchet teeth. When the mainspring has been wound, the energy in the spring causes the barrel and winding arm assembly to rotate and, with the pawl engaged in the ratchet teeth, the ratchet wheel will also turn.

The ratchet wheel is mounted on the spindle of the third wheel,

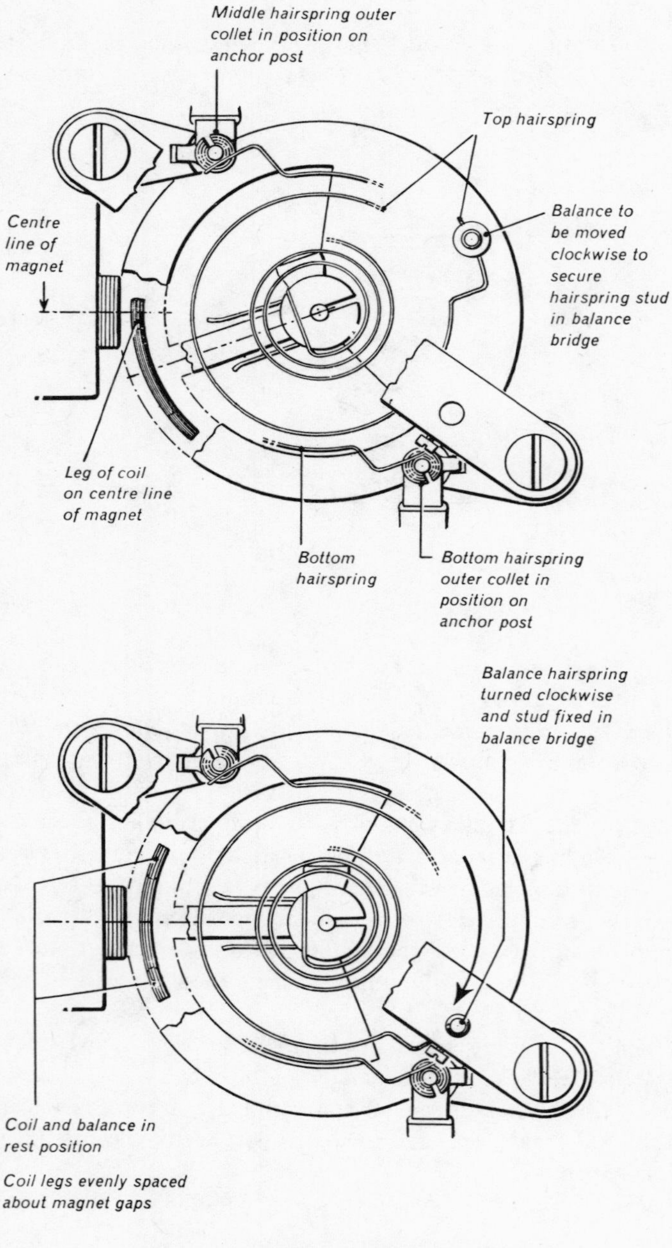

Fig. 153 Securing balance springs

which transmits driving power to the fourth wheel and the escape wheel, where it terminates in a lever escapement. The third wheel also supplies driving power to the wheels that carry the hands.

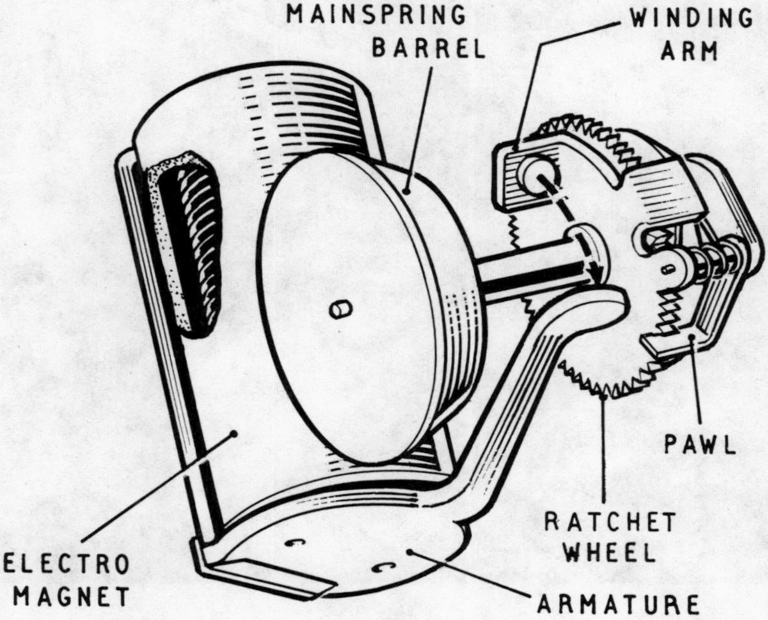

Fig. 154 Kienzle winding arrangement

As the mainspring unwinds, the barrel and the winding arm rotate, bringing the winding arm back to its original position. As soon as the contact on the winding arm touches the contact on the armature, the electric circuit is remade, the electro-magnet is re-energized, and the winding cycle is repeated.

The barrel in which the mainspring is housed is really a balance weight whose center of gravity is eccentric to its axis. As the mainspring runs down, the loss in energy is compensated by the increase in torque produced by the eccentric balance weight. This arrangement provides the movement with constant drive.

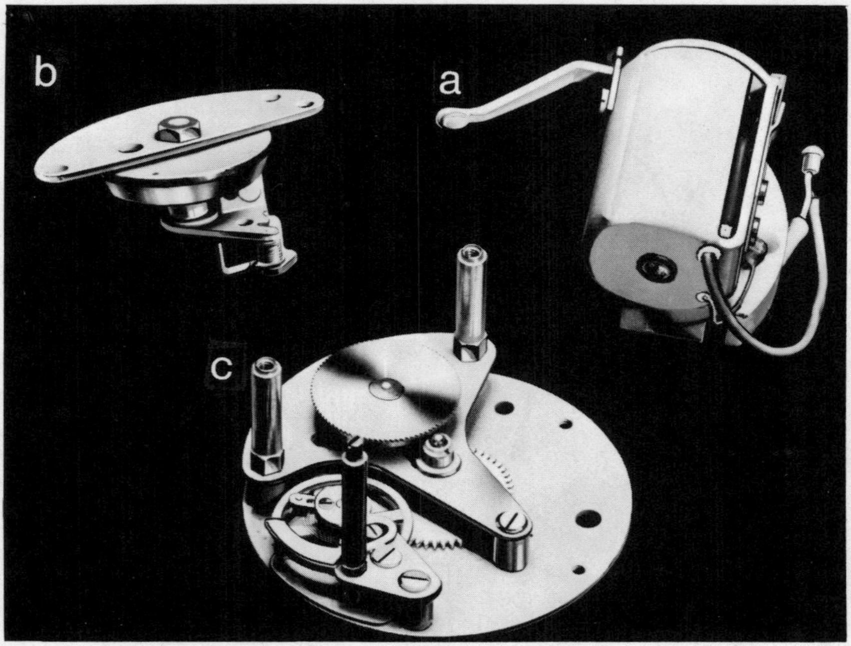

Fig. 155

The battery movement consists of three groups, Fig. 155, which are:

(a) electrical
(b) winding and power storage
(c) mechanical movement

Disassembly is very simple. The fixing screws are removed from the barrel bridge, Fig. 156, and from the insulated base of the electro-magnet, and the two groups are lifted from the movement.

There will be some tension in the mainspring which will cause the barrel to rotate a few turns as soon as the assembly is released from the movement. No further disassembling of these two groups

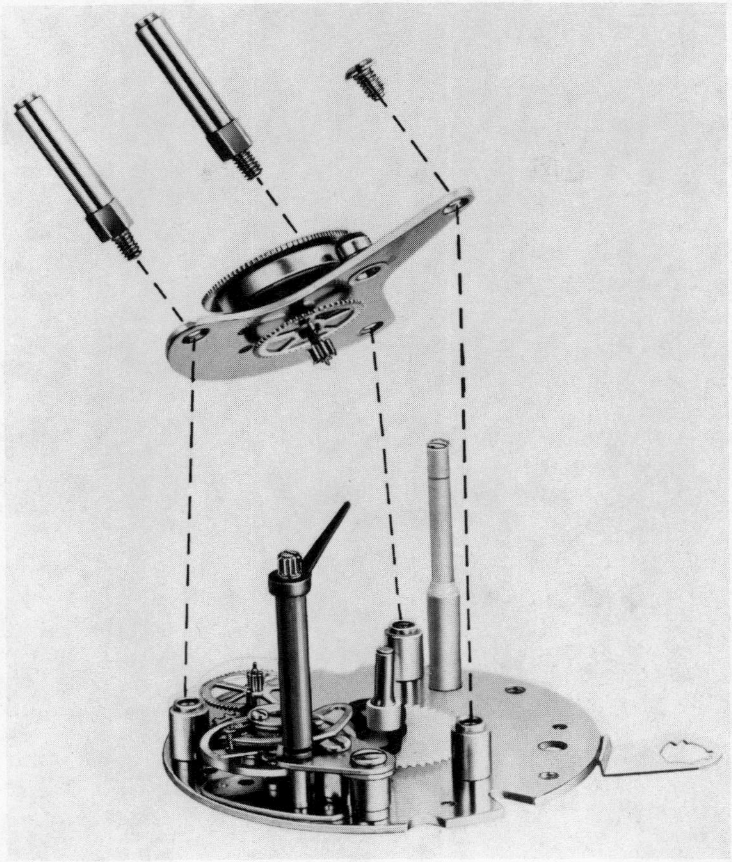

Fig. 156

is required. If the mainspring is broken, the complete winding and power storage unit must be replaced.

To disassemble the mechanical movement, first turn the regulator pinion clockwise until it is disengaged from the pinion rack on the balance cock, and then remove the balance cock screws. The balance can now be lifted from the movement, Fig. 157.

Remove the pallet cock screws and lift out the pallet cock and pallet. Take out the pillar screws, lift off the backplate and remove

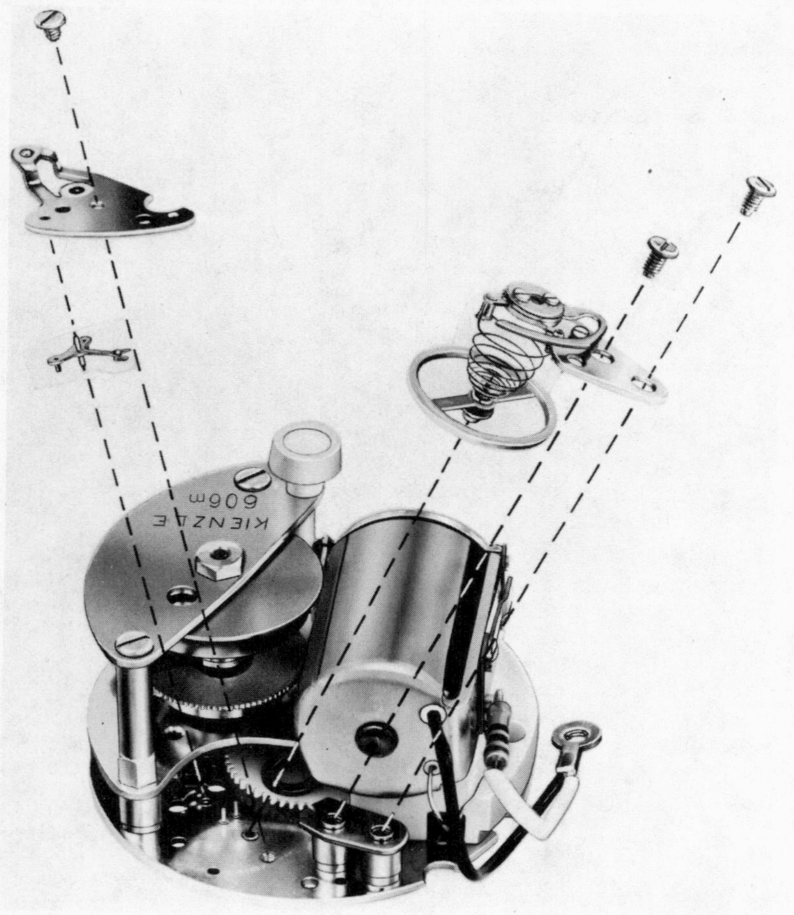

Fig. 157

the wheels. No attempt should be made to remove the third wheel spindle carrying the third wheel and the ratchet wheel from the backplate.

Make sure the wiring of the electrical equipment is in good condition, and that all connections are properly made. Dirt and grease can be removed by wiping with a cloth moistened with CCl_4. No other servicing should be needed. The electrical contact points should never be cleaned with an abrasive and should never be scraped.

The rest of the movement is cleaned, inspected and oiled in the usual way.

Reassembly is straightforward. The wheel train and backplate are fitted first, followed by the pallet and then the balance. When fitting the winding and mechanical power storage unit, make sure that the pawl is engaged in the teeth of the ratchet wheel, and is held there under tension of the pawl spring.

When the electrical unit is fitted it will be necessary to turn the barrel in a counter-clockwise direction to line up the two electrical contact points.

Make sure that the magnet armature has freedom of movement and is not spoiling anything else.

Tension is restored to the mainspring by turning the winding head, Fig. 158, 2½ to 3 turns in a clockwise direction.

Hold the movement vertically, and in an inverted position so that the magnet armature is face up. Note the position of the armature; it should be pressed against the rubber stop on plate 4 by return spring 3. It must not be allowed to rest on coil 2.

If the armature has been kept secure to the coil, and not disturbed, it is unlikely that any adjustment will be needed. However, if there has been any disassembling or renewal of parts, then adjustment may be required. Slacken the two screws 5 and bend the return spring carefully outward. Retighten the screws when the required tension has been obtained.

The distance between the armature and the magnet coil is important but, like the previous check, is hardly likely to be in need of adjustment unless the assembly has been disturbed. As shown in Fig. 158 the distance should be between 0.90 mm and 1.10 mm. Adjustment is made by moving guide plate 4 up or down as required. A simple gauge with which to check the gap is a length of 1 mm diameter wire.

Connect the battery and set the movement in motion. The balance should swing through an arc of between 250 degrees and 270 degrees, and the duration of each winding should be between 90 and 150 seconds. By turning the winding head clockwise, the arc of swing is increased and the time between windings is decreased, and vice-versa.

With any battery-operated clock movement, if the battery has

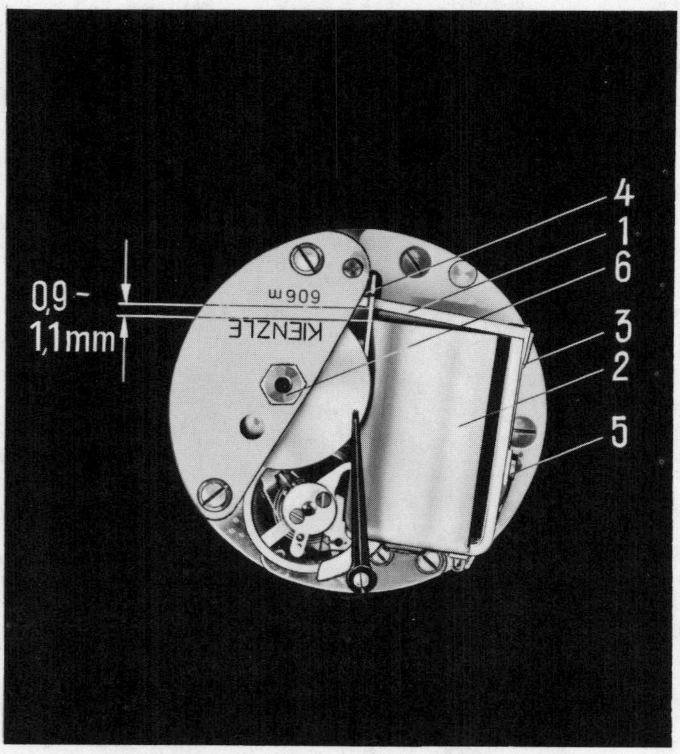

Fig. 158

leaked and has been left in the movement the acid will spread and corrode the metal parts. If the corrosion is widespread, a replacement movement is the best cure. Where corrosion is slight, and confined to a small area, the application of a solution of soap and water will usually clear the trouble.

INDEX